Wilhelm G. Spruth

The Design of a Microprocessor

With Contributions by Members
of the IBM Development Team

With 216 Figures and 8 Color Plates

Springer-Verlag Berlin Heidelberg New York
London Paris Tokyo Hong Kong

Dr.-Ing. Wilhelm G. Spruth
IBM Development Laboratory
Schönaicher Straße 220
D-7030 Böblingen, FRG

The picture on the front cover is a micrograph of the CPU-chip
with a human hair superimposed to indicate the scale.

ISBN 3-540-51395-7 Springer-Verlag Berlin Heidelberg New York
ISBN 0-387-51395-7 Springer-Verlag New York Berlin Heidelberg

Library of Congress Cataloging-in-Publication Data.
Spruth, W. G. The design of a microprocessor / Wilhelm G. Spruth ; with contributions by members of the IBM Development Team. p. cm.
Includes bibliographical references.
ISBN 0-387-51395-7 (U.S.)
1. Microprocessors – Design and construction – Data processing. 2. Logic design. 3. Computer-aided design. I. IBM Development Laboratory (Böblingen, Germany) II. Title.
TK7895.M5S67 1989 621.39'16 – dc20 89-37154 CIP

This work is subject to copyright. All rights are reserved, whether the whole or part of the material is concerned, specifically the rights of translation, reprinting, reuse of illustrations, recitation, broadcasting, reproduction on microfilms or in other ways, and storage in data banks. Duplication of this publication or parts thereof is only permitted under the provisions of the German Copyright Law of September 9, 1965, in its version of June 24, 1985, and a copyright fee must always be paid. Violations fall under the prosecution act of the German Copyright Law.

© Springer-Verlag Berlin Heidelberg 1989
Printed in Germany

The use of registered names, trademarks, etc. in this publication does not imply, even in the absence of a specific statement, that such names are exempt from the relevant protective laws and regulations and therefore free for general use.

Printing: Beltz Offsetdruck, Hemsbach/Bergstraße
Binding: J. Schäffer GmbH & Co. KG, Grünstadt
2145/3140-543210 – Printed on acid-free paper

Preface

This text has been produced for the benefit of students in computer and information science and for experts involved in the design of microprocessors. It deals with the design of complex VLSI chips, specifically of microprocessor chip sets. The aim is on the one hand to provide an overview of the state of the art, and on the other hand to describe specific design know-how. The depth of detail presented goes considerably beyond the level of information usually found in computer science text books. The rapidly developing discipline of designing complex VLSI chips, especially microprocessors, requires a significant extension of the state of the art.

We are observing the genesis of a new engineering discipline, the design and realization of very complex logical structures, and we are obviously only at the beginning. This discipline is still young and immature, alternate concepts are still evolving, and "the best way to do it" is still being explored. Therefore it is not yet possible to describe the different methods in use and to evaluate them. However, the economic impact is significant today, and the heavy investment that companies in the USA, the Far East, and in Europe, are making in generating VLSI design competence is a testimony to the importance this field is expected to have in the future. Staying competitive requires mastering and extending this competence.

The evolving nature of the field makes it impractical to study the individual elements, e.g. logic design, chip design, semiconductor technology design, logical design tools, and physical design tools, in any other way but within the framework of a particular development project, where the pieces fit together. The "Capitol chip set" described in this book has these characteristics. It is a 32 bit microprocessor chip set containing many advanced functions, and represents the work of a design team at the IBM Development Laboratory in Böblingen, Germany. An additional chip, the floating point coprocessor was developed by a team at the IBM Component Development Laboratory in Essonnes, France. The CMOS semiconductor process originated in the IBM Burlington, VT, laboratory.

All contributors to this book are key engineers in the development of the Capitol chip set project. I would like to take this opportunity to thank them for their effort in creating this text, which has been produced largely during off-hours and weekends. The individual contributions have been extensively rewritten in order to create a text of uniform and coherent structure. The contributing authors are individually acknowledged both in the contents listing and at the head of each section written under their technical responsibility.

In addition I would like to thank the remaining 50 or so people who were part of the original project team, as well as those who joined it in the later phases, or supported it in various functions or capabilities, in the laboratories in Böblingen, in Essonnes, in Burlington, VT, and in Yorktown, NY. This was a team of unusually dedicated and competent engineers, and it would be unfair to single out any individuals.

I owe thanks to H. Kircher, Director of the Böblingen Laboratory, to E. Lennemann, Director of System Development, and to V. Goetze, J. Gschwendtner, P. Stollenmeier, J. Thielges, who were members of the Capitol chip set project management team. Special thanks are due to Mrs. H. Kuehnl, Mrs. A. Obermeier, and Mrs. R. Smith for typing, layout, and generation of camera ready output. The layout of the book was performed on an IBM S/370 computer, using Bookmaster as a markup language, Drawmaster as a graphics editor, and an IBM 4250 electroerosion printer together with its postprocessing software, to generate camera ready output. Finally I thank all the staff at Springer-Verlag who were involved in publishing this book.

Böblingen, May 1989   Wilhelm G. Spruth

Contents

Part 1.	**Introduction**	1
1.1.	Overview	1
1.2.	Structure of the Book	2
1.3.	S/370 Architecture	3
1.4.	Layered Processor Structure	4
Part 2.	**Logic Design**	7
2.1.	Design Overview	
	M.Faix	7
2.1.1	Introduction	7
2.1.1.1	Design Goals	8
2.1.1.2	Processor Structure	8
2.1.2	Chip Set Description	10
2.1.2.1	Processing Unit (CPU) Chip	10
2.1.2.2	Memory Management Unit (MMU) Chip	12
2.1.2.3	Storage Controller (STC) Chip	13
2.1.2.4	Clock Chip	14
2.1.2.5	Control Store (CS) Chip	14
2.1.2.6	Floating Point Unit (FPU) Chip	15
2.1.3	Chip Interconnection Busses	16
2.1.4	Cache Timing Considerations	16
2.1.5	Miscellaneous	17
2.1.5.1	Reliability, Availability, Serviceability (RAS)	17
2.1.5.2	System Measurement Interface (SMI)	18
2.1.5.3	Other Processor Components	18
2.2.	Processing Unit Chip	
	D.Bock	19
2.2.1	Design Considerations	19
2.2.2	Block Diagram Description	20
2.2.3	Modes of Operation	21
2.2.3.1	S/370 Mode	24
2.2.3.2	Micromode	25
2.2.3.3	Forced Operations	26
2.2.4	Pipelining	27
2.2.5	Data Local Store Layout	31
2.2.6	Micro Instructions	31
2.2.6.1	Data Local Store Addressing	33
2.2.6.2	Microinstructions Types and Formats	34

2.2.6.3	Reducing the Number of Branch Microinstructions	36
2.2.7	Data Flow Logic	37
2.2.7.1	ALU, Shift, Units and DLS	37
2.2.7.2	Bus Unit and Processor Bus Operations	39
2.2.8	Prefetch Buffer	41
2.2.9	Floating-Point Coprocessor Interface	43
2.3	Timer Support	
	W. Kumpf	44
2.3.1	Introduction to the Timer Functions	44
2.3.2	Format of the Timer Binary Counters	45
2.3.3	Functional Description and Block Diagram	46
2.3.4	Communication with the CPU	48
2.3.5	Programmable Clock Cycle Time	49
2.4.	Memory Management Unit Chip	
	H. Fuhrmann	49
2.4.1	Overview	49
2.4.2	Storage Hierarchy Elements	50
2.4.2.1	Virtual Storage Addressing	50
2.4.2.2	Translation-Lookaside Buffer	51
2.4.2.3	Cache Operation	53
2.4.2.4	Key Store	54
2.4.2.5	Combined Operation	55
2.4.3	MMU Chip Data Flow	56
2.4.4	Array Macros	63
2.4.4.1	Translation-Lookaside Buffer (TLB)	64
2.4.4.2	Cache Directory	67
2.4.4.3	Cache Array	70
2.4.4.4	Keystore	73
2.4.5	Storage Controller Interface	74
2.5.	Storage Controller Chip	
	R. Müller	76
2.5.1	STC Chip Structure	77
2.5.1.1	Memory Control Unit	77
2.5.1.2	Error Correction Unit	79
2.5.1.3	STC Chip Interfaces	80
2.5.2	Memory Organization and Control	
2.5.2.1	Overview	81
2.5.2.2	Memory Card	81
2.5.2.3	Memory Performance	83
2.5.2.4	Fetch Operation	84
2.5.2.5	Store Operation	85
2.5.3	Data Integrity	85
2.5.3.1	Refresh	86
2.5.3.2	ECC (Error Correction Codes)	88
2.5.3.3	Complement Retry	92
2.5.3.4	Redundant Bit	93
2.5.3.5	Scrub	94

2.5.3.6	Address Fault Protection	95
2.5.4	Diagnostics	96
2.5.5	Personalization	97
2.6.	Floating Point Coprocessor	
	J.-L. Peter, C. Sitbon, W. Kumpf	98
2.6.1	General Description	99
2.6.2	Floating-Point Instructions and Data Format	100
2.6.3	FPU Interface and Communication	100
2.6.4	Chip Logical Description	103
2.6.4.1	Overview	103
2.6.4.2	Exponent Dataflow	104
2.6.4.3	Mantissa Dataflow	104
2.6.5	Reliability, Checking and Testing	106
2.6.5.1	Reliability	106
2.6.5.2	Checking	106
2.6.6	Performance	107
2.7.	Bus Interface Chips	108
2.7.1	Overview	108
2.7.2	MBA Chip	109
2.7.3	BCU Chip	110
2.8.	Clock Chip	
	K.-D. Müller, D. Schmunkamp	112
2.8.1	Central Clock Generation Versus Distributed Clock Generation	112
2.8.2	Logical Implementation	113
2.8.3	Timing Tolerances, Reset and Checking	114
2.9.	Clocking	
	H. Schulze-Schoelling	118
2.9.1	Clock Signal Types	118
2.9.2	Clock Generation Flow	118
2.9.3	Clock Pulse Generation	119
2.9.4	Evaluation of the Clock Skews	120
2.9.5	Logic Chip Clock Distribution	122
2.9.5.1	Standard On-Chip Clock Distribution	123
2.9.5.2	High Performance On-Chip Clock Distribution	123
2.9.6	Evaluation of the Clocking Scheme	125
2.9.7	Clock Variation	125
2.10.	Processor Bus	
	H.Fuhrmann, H. Schulze-Schoelling	128
2.10.1	Processor Bus Connections	128
2.10.2	Processor Bus Implementation	128
2.10.3	Processor Bus Operation Example	129
2.11.	Reliability, Availability, Serviceability	
	P. Rudolph	132
2.11.1	Overview	132
2.11.2	RAS Strategy and Requirements	134
2.11.3	Initial Chip Set Start and Loading	135
2.11.4	Error Detection	139

2.11.5	Machine Check Handling	140
2.11.6	Support Interface	141
2.11.6.1	Unit Support Interface Description	141
2.11.6.2	Unit Support Interface Operation	142

Part 3. Logic Design Tools 145

3.1.	Logic Design System Overview	
	W. Rösner	145
3.2.	Hardware Design Language	148
3.2.1	Overview	148
3.2.2	The Design Level	149
3.2.3	Design Rules Checks	151
3.2.4	The Macro Level	154
3.2.5	The System Level	155
3.2.6	Design System Dataflow	155
3.2.7	Overall Comparison with VHDL	156
3.3.	Logic Synthesis	
	B. Kick	158
3.3.1	Overview	158
3.3.2	Logic Synthesis Methodology	158
3.3.3	LSS Overview	160
3.3.4	Technology Information	164
3.3.5	Partitioned Synthesis	167
3.3.6	Synthesis Experience	169
3.4	Logic Synthesis Design Experience	
	H. Kriese	171
3.4.1	Overview	171
3.4.2	The Design System	171
3.4.3	Challenges in Using LSS	172
3.4.4	Delay Optimization and the Use of LSS	173
3.4.5	Results and Designers' Echo	174
3.4.6	Discussion	176
3.4.7	Conclusions	176
3.5.	Timing Analysis and Verification	
	S. Heinkele	177
3.5.1	Overview	177
3.5.2	Delay Equations	177
3.5.3	Capacitance Estimate	178
3.5.4	Multiple Clock Designs	181
3.5.5	Multiple Cycle Paths	182
3.5.6	Global Timing Correction for Logic Synthesis	183
3.6.	Logic Design Verification	
	H. Kriese	184
3.6.1	Overview	184
3.6.2	The Concept of Using Logic Simulation	185
3.6.3	Modelling Requirements	186
3.6.4	The Phases in Logic Design Verification	187

3.6.5	What Drives the Simulation - Testcases	188
3.6.6	The Testcase Execution Control Program	189
3.7.	Logic Simulation	
	W. Rösner	190
3.7.1	Overview	190
3.7.2	Hardware Specification Languages	194
3.7.3	Compilation Techniques	196
3.7.4	Simulation Control	198
3.7.5	Distributed Simulation	199
Part 4.	**Chip Technology**	**201**
4.1	Chip Technology Overview	
	G. Koetzle	201
4.1.1	Technology	202
4.1.2	Circuit Libary and Chip Image	203
4.2.	Master Image Chip	
	H. Schettler	204
4.3.	VLSI Book Library and Array Macros	
	O. Wagner	207
4.3.1	Cell Design	207
4.3.2	Circuit Library	208
4.3.3	Sub-Circuit Elements	209
4.3.4	Macro Design	210
4.4.	A New I/O Driver Circuit	
	T. Ludwig	211
4.4.1	Problem Definition	211
4.4.2	Driver Family	212
4.4.3	Dynamic Control	214
4.5.	Embedded Array Macros	
	K. Helwig, H. Lindner	215
4.5.1	Array Configurations	218
4.5.2	Storage Cell and Circuit Design	220
4.5.3	Array Integration	221
4.5.4	Testing of Embedded Arrays	222
4.6.	Packaging	
	R. Stahl	227
4.6.1	Overview	227
4.6.2	First Level Packaging	227
4.6.3	Electrical Considerations	233
4.6.4	Second Level Package	233
Part 5.	**Semiconductor Technology**	**243**
5.1.	Design for Testability	
	M. Kessler	243
5.1.1	Overview	243
5.1.2	Failure Types and Failure Models	244
5.1.3	Structural Test	244

5.1.4	Design for Testability	245
5.1.5	LSSD (Level Sensitive Scan Design)	246
5.1.5.1	Overview	246
5.1.5.1	LSSD Rules and Partitioning	247
5.1.6	Additional Test Features	248
5.1.6.1.	Internal Tristate Driver	248
5.1.6.2	Observation Points	249
5.1.6.3	Logic Circuit Layout Optimized for Defect Sensitivities	252
5.1.7	Random Pattern Testing	252
5.1.8	Auto Diagnostic	255
5.2.	Test and Characterization	
	J. Riegler, P.H. Roth, D. Wendel	255
5.2.1	Wafer and Module Test	255
5.2.1.1	Test Overview	255
5.2.1.2	Process Parameter Test	257
5.2.1.3	Logic Test on Wafer and Module	258
5.2.1.4	Functional Pattern Test	258
5.2.1.5	Logic Test Equipment	259
5.2.2	Failure Localization and Characterization	259
5.2.2.1	Second Metal Test	259
5.2.2.2	Internal Probing Station	260
5.2.2.3	Fail Locating by Internal Probing	262
5.2.2.4	Performance Verification	262
5.3.	Semiconductor Process / Device Design	
	K.E. Kroell, M. Schwartz, D. Thomas	264
5.3.1	The Semiconductor Process	264
5.3.2	Layout Rules	267
5.3.3	Electrical Device Properties	268
5.4.	Failure Analysis	
	B. Garben	270
5.4.1	Purpose of Failure Analysis	270
5.4.2	Failure Analysis Strategy and Methods	270
5.4.3	Failure Analysis Examples	272
5.4.3.1	Particles	272
5.4.3.2	Metal Interruptions at Steep Steps	272
5.4.3.3	Oxide Residues in Contact Holes	273
5.4.3.4	Leakage between Vdd and Ground	274
5.4.3.5	Signal to Ground Leakage	276
5.4.3.6	Source-Drain Leakage	277
5.4.3.7	Latch-Up	278
Part 6.	**Physical Design Tools**	**281**
6.1	Physical Design Concept	
	G. Koetzle	281
6.2	Hierarchical Physical Design	
	U. Schulz, K. Klein	283
6.2.1	Methodology	283

6.2.2	Partitioning and Floorplanning	283
6.2.3	Implantation	285
6.2.4	Detailed Processing	286
6.2.5	Chip Assembly	291
6.3	Hierarchical Layout and Checking *K. Pollmann, R. Zühlke*	292
6.3.1	Chip Layout	292
6.3.2	Chip Merge and Final Data Generation	292
6.3.3	Checking	293
6.4.	Delay Calculator and Timing Analysis *P.H. Roth, H.G. Bilger*	294
6.4.1	Circuit Delay	294
6.4.2	Calculation Method and Simulation	295
6.4.3	Fitting Method (Least Square Fit)	296
6.5.	Physical Design Experience *G. Koetzle*	298
6.5.1	Master Image Development	298
6.5.2	Physical Design	299
6.5.3	Hardware Bring-Up	299
6.5.4	Lessons Learned	300
Part 7.	**System Implementation**	**301**
7.1.	ES/9370 System Overview	301
7.2.	High Level Microprogramming in I370 *J. Märgner, H. Schwermer*	303
7.2.1	Overview	303
7.2.2	Concepts and Facilities	304
7.2.2.1	Processor Structure	304
7.2.2.2	Instruction Interpretation	305
7.2.2.3	Control Spaces and Associated Instructions	305
7.2.2.4	Mode Control and Associated Instructions	307
7.2.3	ES/9370 Realization	308
7.2.3.1	ES/9370 System Structure	309
7.2.3.2	Service Processor to I/O Controller Communication	310
7.2.3.3	Extending the Kernel Functions	311
7.2.3.4	Simulation Concept for I370 Programs	313
7.2.4	Conclusions and Outlook	314
7.3.	System Bring-Up and Test *W.H. Hehl*	316
7.3.1	Overview	316
7.3.2	Bring-Up Strategy	318
7.3.3	Basic Bring-Up Process	318
7.3.3.1	Sub-Architectural Verification	320
7.3.3.2	Architecture Verification	321
7.3.3.3	Testing Under the PAS Control Program	321
7.3.3.4	System I/O and Interaction Testing	322
7.3.4	System Bring-Up	322

7.3.5	Regression Testing	323
7.3.6	Bring-Up Results and Error Corrections	324
7.3.6.1	Basic Bring-Up	325
7.3.6.2	System Bring-Up	325
7.3.7	Summary and Conclusions	325
7.4.	Outlook	326

Authors 329

References 335

Glossary 337

Index 341

Figures

1. S/370 programming model 4
2. Layered processor structure 5
3. S/370 processor structure using the Capitol Chip Set 9
4. Chip and module parameters of the Capitol Chip Set 10
5. CPU chip overview..20
6. CPU chip block diagram, data flow and bus unit part 22
7. CPU Chip block diagram, control part..23
8. Stage pipeline 28
9. Pipelined instruction execution..29
10. Pipelined μ-instruction execution..32
11. Data local store layout..34
12. ALU microinstruction format 35
13. MOVE instruction layout...36
14. Example for a FOB loop...37
15. Prefetch buffer..42
16. Format of the timer binary counters 46
17. Block diagram of the timer hardware support logic..................47
18. Timeslot allocation scheme..48
19. S/370 virtual address format...50
20. Segment/page table..52
21. Full associative translation-lookaside buffer 53
22. Non-associative translation-lookaside buffer..............................54
23. Imaginary subdivision of a 16 M byte address space into 128 imaginary address spaces of 128 K byte each..............................55
24. 2-way "set associative" translation-lookaside buffer 56
25. Cache concept 57
26. S/370 cache mechanism 58
27. IBM S/370 key store..59
28. Serial access to high speed buffers..60
29. Simultaneous access to TLB, cache, and cache directory........61
30. Parallel buffer access 62
31. MMU chip data flow 63
32. TBL structure...65
33. Cache directory structure 68
34. Cache directory compare logic 69
35. LRU logic schematic...70

36.	Cache structure	71
37.	Keystore structure	74
38.	S/370 storage key	75
39.	Key_status_bus layout	74
40.	Interface to storage controller	76
41.	System structure	77
42.	Memory control unit	78
43.	Command / address format	78
44.	Error correction unit	79
45.	Cache controller interface	81
46.	Main memory interface	82
47.	8 Mbyte memory card organization	83
48.	RAS / CAS access	84
49.	Data transmission vs interleave modes	85
50.	Fetch 64 byte	86
51.	Memory chip organization	87
52.	Store 64 bytes	89
53.	Distance of codes	89
54.	Error correction principle	90
55.	Check bits vs data bits	90
56.	Check bit matrix	90
57.	Single error detection / correction	91
58.	Double error detection	91
59.	Hard error correction by complement retry	93
60.	Double error detection / correction	94
61.	Retry timing	95
62.	Memory organization	96
63.	Refresh / scrub address counter	97
64.	Start-up procedure	98
65.	Different number representations by conversion (example)	101
66.	S/370 floating-point number representation	102
67.	The FPU in the CAPITOL chip set environment	103
68.	FPU chip global data flow	105
69.	Single stage bus structure	109
70.	Dual stage bus structure	110
71.	I/O bus interconnection	111
72.	Hardware I/O buffer area in non-/370 mainstorage	111
73.	Clock chip configuration	113
74.	Clock chip overview	114
75.	Clock macro overview	115
76.	Clock pulse timing	116
77.	Distribution of delay line inputs to selectors	117
78.	Clock signals overview	119
79.	Clock generation flow	120
80.	Clock pulse generation	121
81.	Standard on-chip clock distribution	122
82.	High performance on-chip clock distribution	124

83.	Basic clock triggering	124
84.	C1 and B1 clock setting	126
85.	Clocking scheme for 80 ns cycle time	127
86.	Processor bus	129
87.	Principle of the processor bus driver logic	130
88.	Processor bus timing	132
89.	Repair action learning curve	133
90.	Product quality goal setting	134
91.	SRL chain structure	136
92.	Chip set reset lines	138
93.	Support bus interface	142
94.	Support bus interconnection	143
95.	Major logic design steps	146
96.	Design language example	146
97.	Two alternative ways to generate BDL/S data	147
98.	Logic design and verification overview	147
99.	Design language description layers	148
100.	Simple ALU Design	150
101.	4 bit adder enclosed in shift register latches (SRL's)	152
102.	4-bit adder design language example	153
103.	Physical attribute "AT"	154
104.	Macro hierarchy	154
105.	Example for the system level	156
106.	Design system dataflow	157
107.	Logic synthesis methodology	159
108.	Design language level transforms	161
109.	Decoder/selector transforms	162
110.	Fanin correction	163
111.	Inverter reduction example	164
112.	Timing correction transform	165
113.	Fanin ordering and reduction	166
114.	Outline of fanout correction transform	167
115.	Global fanout correction (max. fanout = 6)	169
116.	Design system	172
117.	Delay distribution	174
118.	Writing capacitance estimate for the CPU chip	180
119.	Writing capacitance estimate for the MMU chip	181
120.	Multiple clock design	182
121.	Iterated timing correction on a path over two partitions	184
122.	Simulation system structure	191
123.	Mixed level simulation methodology	192
124.	Alarm chip	194
125.	Example for behavioral language	195
126.	Action compilation	197
127.	Structure of the simulator	198
128.	Sim Control example	199

129. Concurrency in co-simulation 200
130. ASIC (Application Specific Integrated Circuit) development 201
131. Critical path configuration 202
132. Critical path delay as a function of the level of integration at a
     given packaging technology 203
133. Master image chip cell arrangement 205
134. Distribution of wire netlength of a user representative 50k gates
     MIDS VLSI chip 208
135. Skew compensation in a long logic path 209
136. Layout of a shift register latch 211
137. Driver circuit environment 212
138. Off-chip driver delays 213
139. Driver circuit block diagram 214
140. DISCO circuit block diagram 215
141. MMU chip with embedded memory macros 216
142. Array configurations 216
143. Cache 217
144. Cache directory/TLB 217
145. Key store 219
146. Cache timing diagram 219
147. CMOS 6 device cell 223
148. Keystore sense amplifier 223
149. NOR decoder (for TLB) 224
150. Tree decoder 225
151. Redundancy Steering 226
152. Array integration in logic 226
153. Processor card 228
154. Substrate with double layer metallurgy 229
155. Substrate with single layer metallurgy 230
157. Double layer metallurgy fanout structure with chip connection dots 231
157. Single layer metallurgy fanout structure with chip connection dots 232
158. Detail of double layer structure 233
159. Huffman sequential network 246
160. SRL's shift registers 248
161. Non-overlapping clock structure in LSSD 249
162. LSSD double latch design 250
163. Interface between logic and arrays 250
164. Input to and from array 251
165. Compute time for test generation 251
166. Logic cones form a size partition 252
167. Basic layout of the LSSD chips 253
168. RPT tester environment 254
169. Capitol chip set test strategy 256
170. Channel length spread vs hardware lots 257
171. Adapter board with 256 pin contactor probe 260

172.	Principle of logic path delay measurement	263
173.	Cross section through the various layers of the process	265
174.	Layout rules for transistors and wiring levels	266
175.	Threshold voltage as a function of electrically effective channel length	267
176.	Particle which caused a short between two metal lines	273
177.	Metal interruption at steep steps	274
178.	Silicon dioxide residues in contact holes	275
179.	Leakage between Vdd and ground caused by misalignment between contact holes and polysilicon lines	276
180.	Defective via hole in the polyimide insulation between the 2nd and 3rd metal layer causing a tristate driver fail	277
181.	Transmission electron micrograph of a transistor, which failed because of source-drain leakage	278
182.	Silicon damage caused by latch-up	279
183.	VLSI CAD computer requirements	282
184.	Hierarchical physical design processing flow	284
185.	The sum of all interconnections between the units of logic is continuously lowered until islands-regions-emerge (sum = 0)	285
186.	Positioning of regions on chip based upon the number of interconnections and area assignment	286
187.	Partitioning process steps	287
188.	Partition processing	288
189.	Chip assembly processing	289
190.	Grouping of regions to form partitions	290
191.	Generating interconnect-pins (ic-pins) by pseudo-routing all nets across pre-defined ic-areas	290
192.	HPCMOS chip image elements	293
193.	2-Way NAND circuit	295
194.	Transient behavior of 2-way NAND circuit	296
195.	Input pulse generation	297
196.	Circuit sample	298
197.	Overview of system components	302
198.	Basic IBM ES/9370 processor structure	303
199.	Conventional S/370 processor	305
200.	Microcode analysis without I370	306
201.	A S/370 processor with I370	307
202.	Microcode analysis with I370	308
203.	I370 dispatcher	309
204.	I370 unique instructions	310
205.	I370 control spaces	311
206.	ES/9370 system structure	312
207.	System communication structure	313
208.	I370 support and applications	314
209.	Simulation environment	315
210.	Development phases (schematic) from pilot hardware availability to general product availability	317

211. Bring-up schedule 319
212. Bring-up and system verification levels 320
213. Problem statistics evolution in IBM 9373 Model 30 bring-up time 323
214. Engineering levels of the chips of the Capitol chip set 324
215. Memory chip density 326
216. S/370 processor performance growth 327

Color Plates

1 Master image with horizontal and vertical power lines 235
2 Power distribution on metal 3 236
3a Set of subcircuit elements 237
3b Layout of a 3 way NAND 237
4a First and second level wiring 238
4b 6 device cell 238
5 CPU chip placement 239
6 CPU chip wiring 240
7 MMU chip wiring 241
8 MMU wafer 242

Part 1. Introduction

1.1 Overview

Integrated circuit silicon technology experienced a very fast growth in technological capabilities during the last 20 years. The creation of complex VLSI (Very Large Scale Integration) designs is becoming a major economic factor, and employs an increasing number of engineers and computer scientists. Spearheading this development is the implementation of a "microprocessor" on a single chip.

To a large extend the value of a computer architecture is determined by the amount of system and application software that has been written for it. Therefore, microprocessor implementations have started to appear for all major minicomputer and mainframe architectures. This is for example true for the DEC VAX architecture, with its Microvax II and Microvax III implementations.

The Capitol chip set described in this book is an implementation of an existing mainframe architecture. Due to silicon real estate constraints, most microprocessors do not consist of a single chip, but a chip set. Major components, as a rule, are the basic CPU chip, the memory management unit chip, a floating point coprocessor chip, and a bus adapter chip. The Capitol chip set follows the same pattern. It also includes a separate clock chip, a specialized high speed control store, and a memory controller chip.

The design of a microprocessor involves several different engineering disciplines, especially logic design, circuit design, technology design, tools design, and system design. These individual design disciplines have all been described elsewhere in the literature. However, their interactions are complex. This book describes not only individual designs, but also how the pieces fit together. Due to the complexities involved, and the evolving nature of the state of the art, these interactions of design parts and design tools can only be shown within the framework of an actual development project. Listed below are several textbooks and articles, that may serve as suitable introductions into some of the topics covered. They also contain extensive literature references.

[BAER] and [TANE] are a recommended base for the topics covered in part 2. [AMDA, CASE, and GIFF] are useful as an introduction into the S/370 architecture.

[MEAD] and [HÖRB] offer a good introduction and some alternative views into the topics covered in parts 4, 5, and 6. A number of references are listed within the body of part 3.

[BELL], concentrating on the design of the PDP 11 family of computers, served as a model for this book in organizing a collective of authors to represent their thoughts.

1.2 Structure of the Book

The following text is structured along the lines of the individual engineering disciplines involved in the design of a microprocessor. Part 2 covers logic design. Its sections describe the individual chips, their interconnection, and the clocking scheme. One section is devoted to the important issue of reliability, error diagnosis and recovery.

Part 4 covers the design of the logic circuits and arrays, the chip image, drivers, and packaging. The CMOS technology, its characteristics, testing requirements, and related issues are discussed in part 5.

In the past, engineers used to draw circuit schematics on a piece of paper. They then used this information to manually transform it into a working prototype. The complexities of a modern microprocessor design require a fully automated, computer assisted design process. Since trouble shooting on a VLSI chip is nearly impossible, silicon hardware has to work essentially error free the first time it is produced, without the benefit of traditional trial and error approaches. The Capitol chip set used "logic synthesis", supplemented by a closely related simulation and design verification approach, which is described in part 3. Placement, wiring, and mask data generation steps are covered in part 6.

The final part 7 covers the implementation of a particular system that utilizes the Capitol chip set.

Some of the topics covered in this book may be of special interest:

Modern implementations of existing architectures frequently borrow concepts from Reduced Instruction Set Computers (RISC). This applies in particular to the idea of implementing the less complex (and most frequently used) instructions through hardwired logic, and the remainder in microcode. Section 2.2 covers the resulting dual pipeline, processing microinstructions and hardwired machine level instructions simultaneously.

Section 2.4 explains how accesses to the Translation-Lookaside Buffer, the cache and the cache directory can occur in parallel. Section 2.5 discusses the implementation of error correction and error recovery for dynamic random access main memories. The importance of clock design and clock skew minimization (sections 2.8 and 2.9) is often underestimated, although clock skew contributes a sizeable and growing part to the machine cycle time. Reliability,

Availability and Serviceability features (section 2.11) will certainly grow in importance as the state of the art progresses.

Sections 4.2 and 4.3 discuss an effective approach to implement VLSI logic with a reduced and easily changeable book set. The new I/O driver design in section 4.5 helps to reduce the machine cycle time. Level Sensitive Scan Design (section 5.1) has been around for a long time, but has only recently been used in a growing number of designs outside IBM. The discussion of semiconductor failure mechanisms in section 5.4 is supported by a number of convincing photos.

The combined approach of silicon compilation and design simulation covered in part 3, plus the hierarchical physical design covered in part 6, have not been discussed elsewhere at a comparable level of detail. The high level language microcode implementation of section 7.2 is new, while the system-bringup described in section 7.3 relates to another topic that has been neglected in the literature.

1.3 S/370 Architecture

The S/370 architecture implemented by the Capitol chip set is similar in concept to most other 32 bit microprocessor and mainframe architectures, characterized by using an 8 bit Byte as the basic unit of memory addressing, and implementing multiple 32 bit general registers and a 32 bit data path. It is unique in that it was the first computer family architecture that implemented these concepts. Its programming model is shown in Figure 1.

Modern computer architectures are often characterized as either "Complex Instruction Set" (CISC) or "Reduced Instruction Set" (RISC) Computers. Typical CISC architectures are the DEC VAX, and Motorola 68xxx. Typical RISC architectures are the Sun SPARC, IBM 6150, HP Precision, and Motorola 88000. The IBM S/370 architecture falls inbetween these extremes of complexity. It shares a limited number of addressing modi, and the absence of a stack mechanism with most RISC architectures. It has multiple instruction sizes, and a rich set of supervisor functions similar to most CISC architectures. However, instruction sizes are limited to 2, 4, and 6 Bytes. Due to this, the S/370 architecture exhibists an excellent efficiency in main store (and cache) utilisation.

S/370 Computers as a rule display superior characteristics in terms of reliability, integrity, and error recovery. The S/370 architecture (and the Capitol chip set) include certain architecture features to support these characteristics.

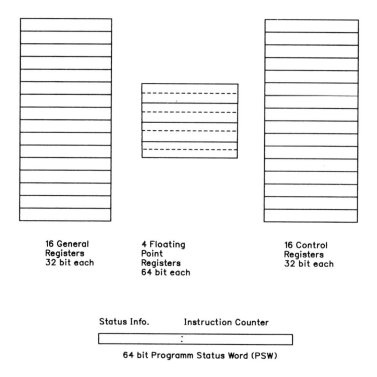

Figure 1. S/370 programming model

1.4 Layered Processor Structure

Except for the RISC, most computer architectures are based on microprogramming. This leads to a layered processor structure, shown in Figure 2. The user interfaces with the computer via Operating System calls. An operating system routine interprets system calls with a sequence of S/370 architecture level machine instructions. A machine instruction in turn is frequently (but not always) interpreted by a microprogram routine, while each microprogram instruction is executed by hardwired controls, usually, but not always, in a single step (machine cycle).

The most frequently used S/370 instructions are not interpreted by microcode, but executed by hardwired controls. Resulting from this, technical details discussed in particular in part 2 frequently apply to processing at either the architecture layer, at the microcode layer, or a combination of layers. For example, the pipeline described in "2.2.4 Pipelining" on page 27 may simultaneous execute a mixture of microinstructions and hardwired S/370 architecture level machine instructions. It may help to frequently consult Figure 2 to clarify at which layer some particular feature operates.

Design Overview

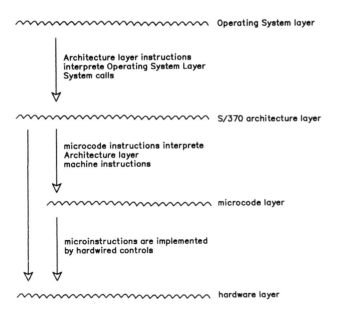

Figure 2. Layered processor structure

Part 2. Logic Design

2.1 Design Overview

Michael Faix

2.1.1 Introduction

The Capitol chip set is a high-performance multi-chip 32-bit microprocessor implementing the System/370 mainframe architecture as described in System/370 Principles of Operation [IBM1]. The implementation uses some structural elements of two predecessor machines, the IBM 4361 and the IBM 9370-90 processors, both realized in high speed bipolar technology. Especially the same microinstruction format is used. The Capitol chip set realization uses a set of 12.7 mm chips in a $1.0 \mu m$. CMOS technology with a high-density master image that contains logic and RAMs. Three layers of metal are used, two for wiring and one for power distribution and I/O redistribution for the central-area pad arrangement. Other chips with a more standard CMOS technology complement the chip set.

The terms "architecture", "implementation", and "realisation" follow the terminology originally proposed by G. Blaauw [BLAA]. Architecture describes what a system performs, its phenomenology. Usually it is defined by the machine instruction set and its surrounding facilities. Implementation is the logical structure which generates the architecture. Where the architecture defines **what** happens, the implementation describes **how** it is made to happen. Realization is the physical structure, which embodies the logical design and is often considered part of the implementation. The realization deals with the **which** and **where** of component selection, allocation, placement and connection.

A Capitol chip set based system may execute the VM/SP, VSE/SP, AIX/370, DPPX/370, MVS/370 operating systems, or any generic operating system using the supported S/370 architecture facilities, e.g. Picks, Mumps, etc.

2.1.1.1 Design Goals

Besides the design rules common in 32-bit microprocessor design the following major design goals were pursued leading to capabilities usually not available in 32-bit microprocessors:

- All logic chips adhere to IBM LSSD design rules (see "5.1.5 LSSD (Level Sensitive Scan Design)" on page 248).

- Reliability, Availability, and Serviceability (RAS) capabilities are similar to those found in IBM S/370 mainframe CPUs. Special features provide extensive check coverage of data and control logic, extensive memory fault tolerance, and an interface on each logic chip to a generalized Support Bus allowing an external device (Service Processor, e.g. a PC with the appropriate adapter card) to access and manipulate internally latched data on each logic chip provided the clock is stopped.

- a System Measurement Interface (SMI) permits the measurement of events via an external measurement device. Dedicated control lines, triggered by hardware circuitry, allow to trace simultaneously occurring events like CPU state, S/370 hardware mode, micro instruction count, fetch from cache, etc, to accumulate timings, e.g. the time the CPU is in the "wait" state, and to trace microprogram generated events.

- In addition to the vertical microinstructions directly interpreted by the CPU hardware, a S/370 like microinstruction interface, called "internal S/370 mode" (I370) simplifies microprogramming for performance uncritical functions. The I370 microcode interface is aimed to significantly improve microcode development productivity and portability. A higher level language can be used to generate I370 code.

2.1.1.2 Processor Structure

The Capitol chip set supports a uni-processor with multiple virtual address spaces of up to 16 MB each, and a maximum real main storage of 16 MB. As shown in Figure 3 the basic chip set consists of six chips. Some chips, e.g. the Bus to Bus Adapter (BBA) chip, may have multiple implementations for different I/O Bus configurations.

The CPU chip provides the basic processing unit function. The Control Store (CS) chip contains an 8K x 18 bit high speed memory for microprograms. Two of these chips may be used. The Memory Management Unit (MMU) chip performs virtual address translation, contains an 8 KByte cache, and a 4 KByte main storage protection key store. The Storage Controller (STC) chip controls the main store. The Clock chip generates central timing signals. The Floating Point Unit (FPU) chip is optional, and increases floating point performance. The Bus to Bus Adapter (BBA) chip interfaces the Processor Bus to an external I/O bus.

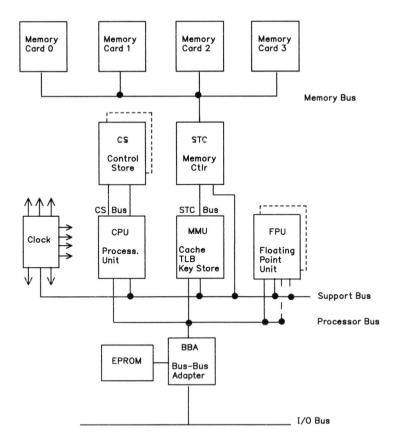

Figure 3. S/370 processor structure using the Capitol Chip Set

The chips communicate with each other via several busses. The Processor Bus is the central interconnection bus for the CPU, MMU, FPU, and BBA chips. The CS Bus interconnects the CPU chip with the CS chip. The STC bus interconnects the MMU and STC chips. The Memory Bus interconnects the STC chip with the memory cards. The Support Bus (USI) provides an interface to an optional Service Processor (SP).

The chips have been designed to operate at a number of different machine cycle times. The IBM ES/9370 system realizations have either a 62.5 ns or an 80 ns cycle.

Figure 4 gives a summary of selected physical and electrical characteristics of the chip and module parameters.

Parameter	CPU	MMU	STC	Clock	CS	FPU
Chip Technology	CMOS (Std. Cell)	CMOS (Std. Cell)	CMOS (Std. Cell)	CMOS (Std. Cell)	CMOS (Array)	CMOS (Std. Cell)
Size (mm)	12.7 x 12.7	12.7 x 12.7	8.6 x 8.6	7.5 x 7.5	7.25 x 6.3	9.4 x 9.4
Transistors	200K	800K	70K	90K		90K
Metalization Levels	3	3	3	2	2	2
Module Pins						
Signal	164	164	164	166	54	150
Power	32	32	32	12	54	24
Power Distribution Active nominal (W)	1.8	1.8	.8	1.8		1.0
Supply Voltage	5V ±10%					
Interface Levels	TTL/CMOS					
Max. Temperature	85°C					

Figure 4. Chip and module parameters of the Capitol Chip Set

2.1.2 Chip Set Description

2.1.2.1 Processing Unit (CPU) Chip

The CPU chip contains the 32-bit data flow including a 32-bit ALU, a 32-bit shift unit, an 8 Byte prefetch buffer for S/370 instructions, and a 4-stage pipeline which allows the overlapped processing of up to 4 S/370 instructions at a time: fetch instructions, decode and fetch operands, perform operation and store result.

The Data Local Store (DLS) holds 16 General Registers, 4 Floating Point Registers, 16 internal Work Registers and 8 internal Status Registers. These 48 32-bit registers can be accessed individually via three separately addressable ports: two output ports and one input port. In addition to the DLS the CPU chip provides hardware facilities for the implementation of the Interval Timer, CPU Timer, Time-of-Day (TOD) Clock and Clock Comparator.

S/370 instructions are fetched by the CPU from S/370 mainstore via the MMU chip and the 32-bit Processor Bus into the 8 Byte S/370 instruction prefetch buffer from where they are executed. S/370 instructions are fetched in advance in increments of one fullword as the prefetch buffer is emptied. In S/370 terminology, full words are 32 bit, half words are 16 bit, and double words are 64 bit wide. During S/370 instruction execution the S/370 Instruction Address Register with its associated modifier updates the address accordingly.

S/370 instruction execution is implemented in one of three ways:

- Based on the observation that the simple S/370 instructions are executed most frequently, seventy S/370 instructions are executed in hardware with no microinstructions involved. Using this subset, the Capitol chip set has RISC like characteristics.

- Vertical microinstructions are used to implement most of the remaining S/370 instructions

- The "internal S/370 mode", the socalled I370 implements some complex S/370 instructions, notably the I/O instructions, and system control functions. Hardware and microcode implement a primitive I/O interface, through the use of "control spaces" which is a variation of the "Memory Mapped I/O" concept applied in many microcomputers.

Vertical microinstructions are 2 Byte wide, most branch instructions, however, are 4 Byte wide. Microinstructions are executed by the CPU hardware. During S/370 instruction execution the CPU chip may fetch microinstructions by placing the Control Store address onto the 13-bit CS Address Bus and reading the microinstruction via the bi-directional 16-bit CS Data Bus at a rate of one per 80 nsec machine cycle.

Microroutines not contained in the Control Store are held in an extension of the main memory not addressable by the S/370 architecture. (non-S/370 memory). The 16-bit Control Store address allows the addressing of 64K 2-Byte microinstructions. When the CPU detects a CS address beyond the largest address implemented in the CS, it issues a microinstruction block fetch operation to the MMU chip, which (with the help of the STC chip) fetches the 64 Byte microinstruction block from the non-S/370 memory area. The CPU stores it into a fixed 64 Byte area in the CS and fetches the microinstructions again for execution. The CS is unaware of the microinstruction paging mechanism. Loading the 64 Byte microinstruction buffer from the non-S/370 memory into the fixed CS area is a time consuming process. Thus only infrequently executed microroutines are stored in this memory. They are organized such as to minimize the number of microinstruction buffer loads while executing microinstruction sequences.

There are two 128 KByte address spaces from which microinstructions can be fetched for execution. The first is the CS address space just discussed. For S/370 systems without a Service Processor the initial microcode supporting the "Initial Microprogram Load" (IML) can be fetched from the EPROM address space. The EPROM is addressed via the BBA chip and the microinstructions are loaded into the CS similar as from the non-/370 memory. Switching from the EPROM address space to the CS address space is microinstruction controlled.

The I370 facilities provide an almost complete S/370 processor with no I/O instructions. I370 instructions execute in the same way as regular S/370 instructions, many of them using the vertical microinstruction set. They share most of the S/370 processor resources, and use a dedicated memory and set of general registers. I370 memory is mapped into the non-S/370 memory. Since the I370 code is invisible to the user it could be considered as microcode, although it employs regular S/370 instructions. For a detailed description see "7.2 High Level Microprogramming in I370" on page 305.

Starting the Capitol chip set from a power off condition requires loading of the microcode and I370 code into the Control Store and the non-/370 memory. Hardware functions on the CPU chip facilitate Initial Microcode Loading via the BBA chip into the Control Store. A checking mechanism ensures proper loading.

2.1.2.2 Memory Management Unit (MMU) Chip

The MMU accepts a real or virtual memory address and reads/writes from/into the cache/memory. It handles S/370 instructions and its data, I/O data, microinstructions and I370 instructions.

All memory requests originating from either the CPU or the FPU chip when executing S/370 CPU instructions, or the BBA when performing I/O operations, are issued against the MMU chip. The MMU performs virtual to real address translation. It provides a Translation-Lookaside Buffer (TLB), an 8 KByte high speed cache and a 4 KByte S/370 key store. As a rule, data are accessed via the cache which supports a store-in-cache algorithm. If data must be accessed in main store, the MMU chip commands the STC chip to do so.

The TLB performs fast translation of virtual addresses into real addresses. Both, the virtual and real page size is 4 KByte. The 64 pre-translated address parts in the TLB are organized in 32 rows with two entries each and are accessed via a 2-way set associative addressing scheme. In case of a TLB miss, the CPU is trapped and a microroutine is invoked which translates the virtual address into a real address by means of the segment and page tables located in S/370 mainstore. A least recently used (LRU) replacement scheme selects the appropriate TLB entry which is then updated to reflect the new virtual and real address parts. A copy of the storage key is fetched from the S/370 Key Store and included into the TLB entry.

An 8 KByte cache with its associated directory provides a high speed buffer that significantly improves processor performance. Cache data and directory arrays are partitioned into 4 columns and addressed via a 4-way set associative addressing scheme. Each column in the cache data array is organized into 32 lines with 64 Byte each. The cache directory holds one entry for each of the 64 Byte lines in the cache data array. In case of a cache miss the MMU automatically sets up a memory command for the STC chip to fetch the required 64 Byte cache line from memory in burst mode. If the cache line to be replaced was changed since it was loaded, the MMU initiates a cache line cast-out operation before the cache line is loaded.

I370 instructions and data are always fed through the cache. 64 Byte blocks of vertical microinstructions fetched from the non-S/370 memory always bypass the cache.

I/O data never causes cache line cast-out and load operations. I/O data to be fetched from main store are first looked up in the cache. If they are not in

Design Overview

cache, they are fetched directly from main store, without cache line replacement. I/O data to be stored into main store are stored into the cache only if the addressed line is already in cache; otherwise, they are stored directly into main store.

The 4 KByte Key Store holds the storage keys for 16 MByte of real memory. It is realized as a 4K x 8 bit array. Each byte holds one storage key. Each TLB entry holds a copy of the storage key associated with the
4 KByte block address to take advantage of the fast TLB access.

Most S/370 main store operations for I/O data cause also the automatic addressing of the S/370 key store. The key read out is compared against the access key and the operation type. If the key matches, normal processing continues; otherwise, an access exception is indicated in an I/O interrupt.

The physical memory (main memory) is between 1 and 16 MByte in size and logically divided into two distinct address spaces: S/370 main store and non-S/370 memory. Both memories are addressed via 24-bit real addresses. The non-S/370 memory holds microroutines, I370 code, internal buffer space, and control data. S/370 main store is mapped into the lower part of the memory. The size of the non-S/370 memory is dependent on the system requirements and may vary between different system implementations. The S/370 main store size is the total memory size installed minus the non-S/370 memory size.

The MMU chip executes the control of the internal memory boundary which separates S/370 main store from non-S/370 memory as well as the S/370 address compare functions.

2.1.2.3 Storage Controller (STC) Chip

The Storage Controller (STC) chip manages the accesses to the dynamic random access memory implemented with either 256K-bit or 1M-bit memory chips. The STC interfaces to the memory via the synchronous bi-directional memory bus that consists of a 32-bit bi-directional data bus plus additional control lines. Up to 4 memory cards are driven by the STC chip.

Memory is assumed to be organized in independent operating banks to allow 2-way or 4-way interleaving. Each 1MByte of memory is treated as a memory bank. Memory Interleaving is utilized when transferring 64 Byte blocks of data from or into the memory. Within a fetch or store 64 Byte block operation, the STC chip controls up to 4 memory banks running in interleave mode. After the first access the memory operates at a data rate of 4 bytes in 80 nsec with 4-way interleaving, and 160 nsec with 2-way interleaving. All memory configurations with at least 4 MByte operate with 4-way memory interleaving. A special "Fetch 64 Byte Slow" operation mode supports loading of 64 Byte microinstruction blocks from the non-S/370 memory into the Control Store. This command operates at a speed of 4 Byte in 160 nsec, because the interface to the Control Store is only 2 Byte wide.

In addition to controlling the basic read and write memory operations the STC chip performs functions like memory refresh, Error Checking and Correction (ECC), scrubbing, redundancy bit control, and others.

2.1.2.4 Clock Chip

The Clock chip generates centrally all the clock signals required by the other chips in the chip set including the BBA. It requires an external hybrid oscillator and an active delay line. It contains multiple clocks. The timing of each clock is independently programmable in terms of pulse width, phase, and cycle time, and thus can be adapted to the varying system requirements. The clock signals feed directly the receiving chips without any further gating. All clock lines are point to point nets, and clock skews are minimized. Clocks are checked within the Clock chip. The Clock chip also provides functions for system power-on recognition, reset and run control. Finally a Support Bus interface is provided for diagnostic purposes.

On power-on recognition the Clock chip is initialized and clock signal generation starts. The clock signals to be delivered at certain reset and start up points are controlled by the run control.

The Clock chip is the master of the chip set for the reset function.

2.1.2.5 Control Store (CS) Chip

The CS chip contains an 8K x 18 bit (147,456 bits) static random access array memory. It stores vertical microinstructions that interpret frequently executed S/370 instructions. The fetching of a 2 Byte vertical microinstruction is completely overlapped with the execution of the current microinstruction. An execution rate of one microinstruction per machine cycle (80 ns) can be maintained. A microinstruction is accessed by the CPU chip from the Control Store by placing a 13-bit CS address onto the uni-directional 13-bit CS Address Bus. The addressed microinstruction is placed onto the 16-bit bi-directional CS Data Bus and stored into the Microinstruction Register for execution. This operation takes one machine cycle. For (infrequent) 4 Byte microinstructions a second machine cycle is required to fetch the second half of the microinstruction. The Control Store access time is approximately 30 nsec, and the minimum cycle time is 40 nsec.

The CS chip is packaged into either a single chip module with 8K x 18 bit, or a dual chip module with 16K x 18 bit. Microinstructions that cannot be contained in the CS are held in the non-S/370 memory.

The CS array has separate but dottable data I/Os. Each off-chip driver contains a latch to hold the data during subsequent chip operations or until another read cycle is completed. The output drivers of the off-chip driver may also be tri-stated without loss of the data in the data latch. The CS chip contains logic that controls the chip functions. Some of the functions available are halfword read/write, decoded read/write operations, and delayed write.

2.1.2.6 Floating Point Unit (FPU) Chip

The CPU chip in conjunction with control store microcode can execute the entire S/370 floating point instruction set. The optional FPU chip provides enhanced performance for all S/370 floating-point, ACRITH [IBM2], and SQRT [IBM3] instructions. The FPU chip acts as a co-processor to the CPU chip, is hardware controlled and operates at a 80 nsec cycle time.

The FPU chip contains a duplicate of the 4 x 64-bit S/370 floating point registers on the CPU chip. It uses a 9-bit ALU for exponent arithmetic, a 62-bit ALU for fraction arithmetic, and an 8 x 56 array multiplier to speed up multiply operations.

During floating point instruction execution the CPU chip decodes and executes the instruction phases (I-phases). Subsequently the S/370 instruction op-code, the floating point register number(s) and the rounding bits for ACRITH instructions are transferred as a message via the Processor Bus to the FPU chip.

Some dedicated bus lines between the CPU and FPU chips speed up communication. The CPU chip uses special commands on the Processor Bus to send status information to and read error information from the FPU chip.

The FPU chip implements the Execution phase (E-phase) of the S/370 floating point instruction set (51 instructions), the S/370 square root instructions (SQER, and SQDR), and some ACRITH instructions by hardware without microinstruction involvement. The remaining ACRITH instruction E-phases are either assisted by or completely executed by CPU microinstructions. The condition code, if any, is sent from the FPU to the CPU and loaded into the current PSW.

Either one or two FPU chips can be attached to the Processor Bus, depending on the level of checking required. Limited checking is available with one FPU chip. At the beginning of a S/370 instruction execution, the FPU checks the validity of the command received from the CPU. Data received from the Processor Bus is checked for correct byte parity, and for data transmitted onto the Processor Bus the appropriate parity bits are generated.

Complete checking is available with two identical FPU chips running in parallel. One FPU chip runs as the master chip. Its output on the Processor Bus is compared against the output of the second FPU chip (shadow approach). Intermediate results (which are not placed on the Processor Bus) are communicated via dedicated lines from the master chip to the slave chip. Checking is always performed on the slave chip which also signals an error condition to either the CPU or MMU chip.

2.1.3 Chip Interconnection Busses

The chips communicate with each other via several interconnection busses:

The Processor Bus is the central interconnection bus for the CPU, MMU, FPU and BBA chips and consists of a 32-bit bi-directional data bus (plus 4 parity bits), a 5-bit bi-directional key/status bus (plus 1 parity bit), and control lines. It is a parallel dotted bus, and synchronously clocked by a two-phase clock. These clocks are distributed to each chip attached to the Processor Bus in an equidistant manner. Each chip can use additional clocks for internal timing requirements. The Bus operates in a clocked, handshake mode unter the control of an arbiter function on the BBA chip. Its design permits an easy interface to the IBM ES/9370 I/O bus, PS/2 microchannel, and industry standard busses.

The Control Store Bus interconnects the CPU chip and Control Store chip(s). The CPU uses the synchronously clocked 13-bit uni-directional address bus to transfer CS addresses to the CS chip(s). The synchronous 16-bit bi-directional data bus (plus 2 parity bits) is used for transferring microinstructions.

The STC Bus interconnects the MMU and STC chips. It consists of a synchronously clocked 32-bit bi-directional data bus (plus 4 parity bits) used for both address and data transfer, and additional control lines.

The Memory Bus interconnects the STC chip and the memory cards. It consists of a synchronously clocked 32-bit bi-directional data bus (plus 4 parity bits) used for address and data transfer, 7 check and redundancy lines and 43 control lines.

The Support Bus (USI) is a 5-wire serial interface that provides a backdoor entry for optional use of a Service Processor , which can have access to the internally latched data in all logic chips. When the clock is stopped the SP can read internally latched data in the logic chips during the execution of micro diagnostics, inject error information during testing, and set up information during the start up phase for loading the microprogram (IML) from an external device.

2.1.4 Cache Timing Considerations

S/370 instructions and data are buffered in an 8 KByte high speed cache on the MMU chip. It was obvious from the design start that the MMU read and write operations determine the basic machine cycle time. The Capitol chip set requires for each S/370 instruction execution approximately two accesses to the cache, one for the instruction itself and the other for data. Due to I/O pin limitations on the CPU chip, no separate 32-bit address bus was implemented, thus address and data are multiplexed on the Processor Bus.

Two design options were investigated:

A **Single Cycle Cache** requires an increase of the basic machine cycle time to allow a full MMU read or write operation within one cycle. The basic machine cycle time, determined by the CPU's ALU operation, must be extended for the time needed to transfer the effective main store address from the CPU chip to the MMU chip and to latch it up. S/370 instruction prefetching can be completely overlapped with the execution of the current S/370 instruction, because the first cycle in each S/370 instruction execution is always a non-cache cycle and the CPU can utilize this cycle for prefetching 4 bytes of the S/370 instruction stream from the cache into its S/370 prefetch buffer.

A **Two Cycle Cache** requires the break down of the cache read or write operation into two cycles, thus leaving the basic machine cycle time independent of the MMU operation. For S/370 instruction prefetching the first non-cache cycle in each S/370 instruction execution is also available to initiate a S/370 instruction prefetch operation from cache. A second non-cache cycle is available with S/370 instructions which require indexing during the effective address calculation or which are executed by microinstructions. In these cases S/370 instruction fetching is also completely overlapped with the S/370 instruction execution.

Performance calculations favored the two cycle cache although its control logic is more complex than for the single cycle cache. Thus the time needed to read/write 4 Bytes from/into the cache is 160 ns for an 80 ns cycle.

2.1.5 Miscellaneous

2.1.5.1 Reliability, Availability, Serviceability (RAS)

Different from commercial microprocessors like the INTEL 80386 or the Motorola 68030, the Capitol chip set offers the same high RAS capabilities available in IBM S/370 mainframe CPUs. To achieve this objective the following features are incorporated:

- Data flow, busses, arrays and memories are parity checked on a Byte basis

- Expanded memory fault tolerance is achieved through ECC (single error correction, double error detection), double error retry for one hard and one soft error, scrubbing to avoid accumulation of soft errors, and fault triggered bank selective automatic redundant bit switching.

- Extensive check coverage of data and control logic is obtained through a checker concept including checking the checkers. Optimum function security is guaranteed by keeping the checkers always active even when executing critical start/restart functions, such as hardware reset, initialization and machine check handling.

- A delayed restart capability is activated after a machine check has occurred to overcome intermittent faults.

- A hardware supported microprogram load provides microprogram bootstrap loading and verification during the load process.

- A Support Bus Interface (Unit Support Bus, USI) is provided on on each logic chip. Each latch in a chip is designed according to LSSD (see "5.1.5 LSSD (Level Sensitive Scan Design)" on page 248) design rules and thus consists of a master and slave latch. All latches in a chip are chained together and are connected (via the USI interface) to the 5-wire Support Bus (with two data and three control lines) which represents a "backdoor" entry to the logic on VLSI chips. An external device (e.g. the PC based operator console that is part of an IBM ES/9370 system) can be used to access and manipulate internally latched data on each logic chip via the Support Bus interface provided the clock is stopped.

- Special circuits on the individual chips facilitate implementation of an operator panel control facility, including error code display and load path selection.

2.1.5.2 System Measurement Interface (SMI)

The SMI consists of a set of control lines for system measurement purposes via an external measurement device.

16 dedicated control lines are triggered by hardware circuitry and available for

- tracing simultaneously occurring events like CPU in problem state (wait state or stopped state), S/370 hardware mode, micro instruction count, S/370 instruction count, fetch from cache, store into cache, cache line store and fetch, cache line fetch, cache line store, TLB update, etc.

- accumulating timings, e.g. the time the CPU is in the "wait" state

Four of the 16 control lines are used for tracing 16 exclusively occurring events generated by the microprogram. All control lines are provided by the CPU and MMU chips.

2.1.5.3 Other Processor Components

Besides the chips described other components, are required to implement a complete S/370 processor. A hybrid oscillator and a delay line are required to support the Clock chip. Resistors are required to terminate all data, control and clock lines. Several BBA chips have been developed for connecting the Processor Bus to IBM ES/9370 I/O busses. A variety of memory cards use either 256-kbit chips or 1-Mbit chips. Storage capacity on one memory card ranges from 4 Mbyte to 16 Mbyte.

2.2 Processing Unit Chip

Dietrich Bock

2.2.1 Design Considerations

The processing unit (CPU) chip is the center piece of the Capitol Chip Set. Its implementation is very similar to the IBM 4361 and IBM 9370-90 CPUs. We know from past experience, that these implementations feature a very good trade-off between performance and number of circuits and arrays employed. However, they are bipolar realisations. In order to translate them into a CMOS realisation, and accommodate the design within the restrictions of the 12.7 x 12.7 millimeter chips and the corresponding packaging technology, a significant redesign activity was required.

The CPU chip works in close cooperation with the MMU chip. Data transfer between the chips may be 1, 2, or 4 Bytes wide. If the operand crosses word boundaries, the transfer time is extended until the CPU chip receives all the data.

Design trade-off considerations, among others a limitation in the number of pins on the CPU chip module forced a two cycle cache access design instead of the one cycle cache access design in the 9370-90 processor. Addresses and data are time multiplexed on the Processor Bus. Command and address information are put on the processor bus in the first cycle, while data are sent in the second cycle.

All existing S/370 processor designs work within an operator console / support processor environment. The Capitol Chip Set has been designed to work without a support processor. For this, several features, such as a separate EPROM address space, the ability to write and read its own control store, and the ability to modify its own configurations (control store size and co-processor installation) have been implemented.

The Capitol chip set executes a machine instruction within one or several units of time, depending on the type of instruction being executed. These units of time are called the "machine cycle". The chips have been designed to run within a number of different machine cycle times, although the existing semiconductor realisation limits them to an 80 ns cycle. For this purpose, the timer facilities were designed such that they can be incremented within a system clock cycle to make timer operation synchronous to all other operations on the CPU chip.

2.2.2 Block Diagram Description

Figure 5 shows an overview of the Processing Unit (CPU) chip. It contains 3 parts, data flow, bus unit and control. A more detailed diagram is shown in Figure 6 and Figure 7. The data flow elements, especially the ALU section with shift unit (SU) and Data Local Store (DLS), the Bus Unit as a link to the Processor Bus, and the Timer, which supports the /370 architectured timer facilities, are connected via one of the three on-chip busses, the four byte wide Central Data Bus.

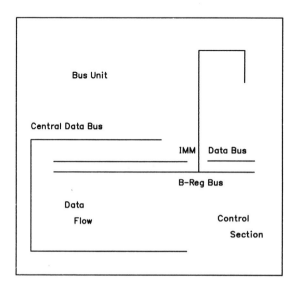

Figure 5. CPU chip overview

The other busses, the Immediate Data Bus and B-Reg Bus perform the data connection from and to the control section on the right side of Figure 5. These buses are unidirectional to ease testing; each bus is two Bytes wide.

The control section splits into 3 sets of logic:

- control of the dataflow directly out of the S/370 Operation Registers for a reduced instruction set of 70 S/370 instructions without microinstruction involvement. This part includes the 370 OP1 REG and 370 OP2 REG in the lower right quadrant of Figure 7 plus their Operation Decoders.

- microinstruction control of the dataflow for implementation of the S/370 complex instructions and control purposes. For this the microinstruction operation registers, the μOP1 REG and the μOP2 REG, and the corresponding Operation Decoders are used.

- control for Forced Operations (FOP), which respond to exceptional conditions in the execution of S/370 instructions and perform transitions between S/370 controlled and microcode controlled operations.

The logical structure of the control section, as shown on the Block diagram, differs from the physical on-chip structure. The Operation Registers are implemented once near the center of the chip. Each operation Register is connected to multiple decoders (Operation Decoders) that are placed close to the dataflow elements. Thus a decoder for a certain type of instruction is implemented more than once. This permits to compare and check the outputs of decoders.

The CPU chip can address these arrays:

- the Data Local Store (DLS), which is located on the CPU chip,
- the Control Store (CS), which is shown amid the control section in the upper half of Figure 7 but is located on a module of its own. The Control Store contains only part of the total functional CPU microcode of the system. Microcode, which does not fit into the control store, is kept in non-S/370 memory.
- The EPROM, which is an additional microinstruction store. It is accessible via the Processor Bus. The microcode in the EPROM can be loaded into the control store and the non-S/370 memory during the Power-on/initialization phase and by error recovery routines.
- The main memory consisting of the S/370 main store and non-S/370 memory is also accessible via the Processor Bus.

2.2.3 Modes of Operation

Logically, the execution of each S/370 instruction splits into an Instruction Phase (I-Phase), which completes the instruction fetch and calculates operand addresses (if any), and an Execution Phase (E-Phase), which manipulates the operand(s). With regard to actual execution, the I-Phase and E-Phase may overlap.

In each machine cycle the CPU chip operates in one of four modes:

- microinstruction mode, also referred to as micromode. This mode is invoked during the Initial Micro Program Load (IML) phase, and during complex S/370 Execution Phases (E-Phase). Exceptional conditions are also handled by microinstructions in this mode.
- S/370 mode, also called hardware mode. All S/370 and I370 instructions start in this mode. They may stay in this mode during the execution phase (E-Phase) or may switch to micromode.
- FOP (forced Operation) mode is the transition from micromode to hardware mode. It is required to load the eight-byte S/370 Instruction prefetch buffer prior to the execution of the first instruction out of that buffer.

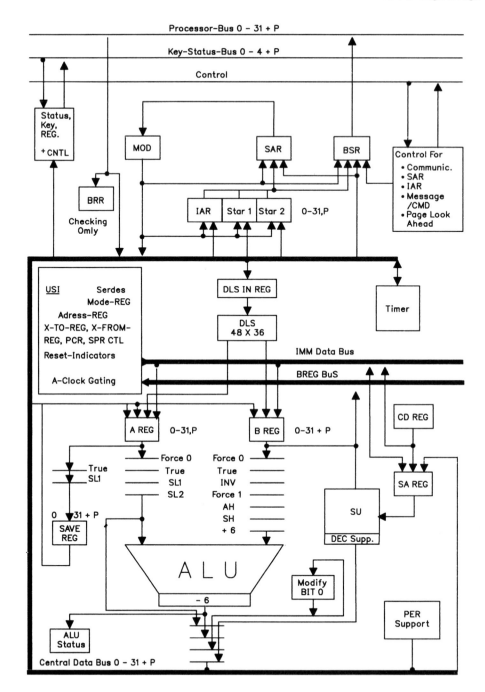

Figure 6. CPU chip block diagram, data flow and bus unit part

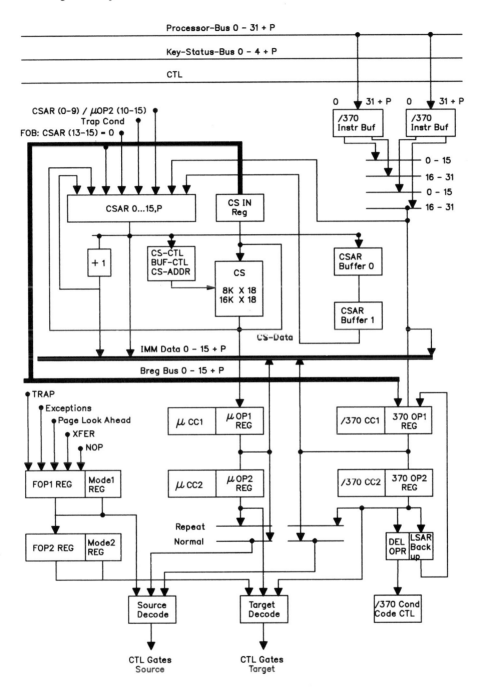

Figure 7. CPU Chip block diagram, control part

- In the FOP mode the CPU-chip reacts to exceptional conditions, which may invoke microinstructions for support.

The four modes are encoded in 2 mode bits and distributed to the operation decoders on the chip. Any decode operation, which activates a gate or sets a latch is an AND function of

- a decode of these mode bits,
- a decode of an Operation Register, dedicated to this mode (e.g. S/370), and
- a cycle of the cycle counter dedicated to this mode, and this stage (for instructions which require two or more machine cycles for execution, see the upper right corner of Figure 7).

The CPU chip implements a 4 stage pipeline. Each processing stage has its own set of mode bits. Normally, a mode transition in the target stage of the pipeline becomes effective one cycle later than in the source stage, but Forced Operations may force mode transitions for source and target stage in the same cycle. That way, a cycle or instruction, which should not been started due to a latent exceptional condition gets a source cycle, but does not generate any results, because its target cycle is suppressed.

The I370 mode (Section "7.3 System Bring-Up and Test" on page 318) does not require any unique hardware facilities except some space in the Data Local Store and in the non-S/370 memory. Transition from S/370 to I370 and vice versa is controlled by microinstructions.

2.2.3.1 S/370 Mode

Each S/370 instruction starts in this mode. The first halfword of the instruction is transferred from the prefetch buffer (Section "2.2.8 Prefetch Buffer" on page 41) to the S/370 Op1 Register. Some S/370 RR format instructions take only one cycle for execution. For all non-RR format instructions the base/displacement address calculation starts in the first cycle. The address calculation overlaps with an instruction prefetch to reload the prefetch buffer. The prefetch buffer reload takes two cycles. Address calculation with indexing (nonzero X2-value) does not require an extra cycle.

S/370 RX and RS format instructions, which do not need data from storage (e.g. Load Address, Shift Left/Right Logical) are completed in two cycles in hardware mode. A mode transition to micromode may or may not occur after address calculation and completion of the prefetch buffer reload. The decision about the mode transition depends on a table, which is located in the first 256 locations of the Control Store. During the same cycle the S/370 Op1 Register is set, the S/370 Operation Code (the first Byte of the instruction) is loaded into the Control Store Address Register (CSAR, upper left quadrant of Figure 7), while the high order Byte of the CSAR is forced to zero. The thus addressed Control Store word contains a pointer to the first microinstruction of a microcode routine that implements the execution of a microcoded E-phase. The pointer thus obtained is loaded into the CSAR. The most significant bit of the pointer controls the transition to micromode; with this bit set to 0 the mode switches, otherwise the CPU stays in S/370 mode.

Whenever micromode gets control for an execution phase the /370 instruction operand storage address or addresses (two for SS format instructions) are already calculated and available in 2 registers. These are the Bus Unit hardware register(s) STAR1 (and STAR2), which are gated to the Processor Bus during the command cycle of the appropriate microinstructions, and one (two) DLS register(s).

2.2.3.2 Micromode

Microinstructions are used to provide a variety of functions: implementation of the execution phases (E-Phases) for complex S/370 instructions, fast response to complex exceptional conditions, IML, and basic test and recovery routines.

Micro instructions run most efficiently, when they execute out of the Control Store. However, only a part of the microcode fits into the control store. The CPU design supports either 8K or 16K halfword (16 bit) control store sizes. Therefore the Control Store locations are reserved for the most performance critical part of the microcode; the remainder is kept in non-S/370 memory. Microcode cannot be executed out of the non-S/370 memory, it has to be loaded in blocks of 64 Bytes into the "Control Store Buffer", an area within the Control Store. The load is performed by the Forced Operation "Transfer". A Control Store Buffer load is initialized, if control branches to an address beyond the maximum address of control store installed.

The Control Store Buffer is the address range from 8160 to 8191 (X'1FE0 to 1FFF'). The function of that address range is controlled by logic on the CPU chip. When assembling and linking the microcode, this address range is left unoccupied.

Microinstructions operate in one of two address spaces, each 64K instructions (128 K Byte) in size:

- the control store address space starts in the control store and extends into the non-S/370 memory.
- the EPROM address space is accessed via the Bus-to-Bus Adapter (BBA) chip. The EPROM address space is used during power-on and Initial Microprogram Load, but not for functional operation.

The transition between the two address spaces occurs with the help of a CONTROL microinstruction, which sets or resets the "EPROM" latch. The CPU chip starts its operation immediately after power-on with a transfer of 64 Bytes of data from the EPROM address space at address X'0000' into the Control Store Buffer, and subsequent execution of this code.

There are 2 sublevels in micromode. Normally, microcode execution in micromode occurs in the "base level". The "trap level" is entered upon detection of a trap condition. The two levels differ by the usage of different internal working registers in the DLS.

2.2.3.3 Forced Operations

The Forced Operations in FOP mode are caused by:

- Traps for unusual conditions from Processor Bus operations and from invalid decimal data,
- Exceptions, mostly S/370 architected exceptional conditions,
- Page Look Ahead for S/370 operands crossing the page boundary.
- Transfer (XFER), which loads microinstructions from main store or EPROM into the 64-byte Control Store Buffer.

There are five causes for a Trap:

- Access to the S/370 store, causing a "miss" of the Translation-Lookaside Buffer, (TLB), (see "2.4.2.2 Translation-Lookaside Buffer" on page 51). Cache misses do not cause traps or exceptions in the CPU chip, but are handled by the MMU chip.
- access beyond the S/370 main store address limit
- protection check during access to the S/370 main store due to a storage key mismatch
- Any Check during a Processor-Bus operation due to malfunction in the participating Processor Bus unit, e.g. parity check or TLB double match
- invalid decimal data in an ALU operation.

Traps and exceptions force transition to micromode via a table in the Control Store, which is located adjacent to the table for the S/370 Operation Codes on addresses 256 to 271 (X'0100 to 010F'). More than five entries are required, because trap conditions are further distinguished for TLB misses with virtual address translation (dynamic address translation, DAT) being ON or OFF and for simultaneous occurrence of more than one trap condition. The content of Control Store locations 256 to 271 is treated as a pointer to the first microinstruction of the corresponding trap routine. Since each trap condition is vectored to a dedicated routine, no condition testing for the cause is required. This especially accelerates time critical TLB miss routines. For all trap causes resulting from Processor Bus operations, the failing address is put into a data local store (DLS) register during the Forced Operation cycle.

The Forced Operation "Trap" activates the trap level (as opposed to base level) for the microcode routine. The effect of the traplevel is threefold:

- a different block of DLS workregister, the trap registers, are used. The execution phase (E-Phase), in which the trap condition occurred, has already set intermediate results into work registers. These values must stay unchanged because the execution of the E-Phase will normally be resumed after handling of the trap condition.

- the change of hardware elements for the S/370 mode (Op Registers and cycle counters) is inhibited.
- a change of CSAR (Control Store Address Register) buffers 0 and 1 is inhibited. CSAR Buffer 1 contains the address of the instruction, which caused the trap.

The trap level is deactivated by a microinstruction at the end of that routine. After a successful resolution of a TLB miss, control returns to the mode of operation from which the trap originated (either micromode or hardware mode). The access, which caused the trap, is repeated.

The Forced Operation "Page Look Ahead" (PLA) is implemented to meet a S/370 architecture requirement: the execution of a S/370 instruction may not be started unless main storage operand(s) are completely accessible in both virtual and real storage. Page Look Ahead is required for operands crossing a 4k page boundary. The crossing condition is detected by a hardware function, which adds the length of an operand, just accessed, to the start address. The operand length is either predetermined for a specific S/370 instruction type, or encoded in the second Byte of the S/370 instruction format.

The addition of start address and length is only done for the first access of an operand. A Page Look Ahead operation is forced, if the result of that addition crosses a page boundary. It forces

- cancellation of the access just started in the next cycle, (the MMU chip may still deliver the data requested together with status information).
- a read access for one Byte with the start address incremented by 4K. This access is also cancelled, so it does not cause a line transfer from main storage to the cache, but may cause a trap to resolve a TLB miss, if a translation is not available in the TLB.
- repetition of the original access.

2.2.4 Pipelining

In hardware or micromode the CPU chip executes instructions in a four stage pipeline.

Many modern processor designs feature a 4-stage pipeline, executing 4 instructions simultaneously. A simplified data flow is shown in Figure 8. Instructions are executed in 4 steps:

- read the next instruction from main store, cache, or prefetch buffer into an operation register

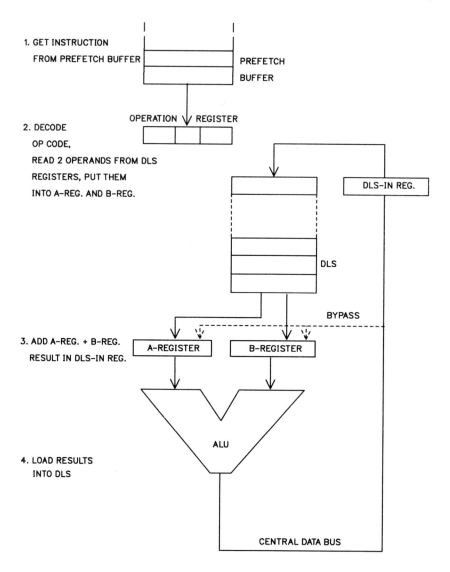

Figure 8. Stage pipeline

- Decode the instruction, fetch the operands (e.g. from the Data Local Store), and load them into the input registers for the ALU/Shift unit, the A and B registers
- Perform the operation, e.g. a binary add, and put the result into a result register (DLS-IN register)
- Store the result register content into the Data Local Store (DLS).

Under optimum conditions, each step takes 1 machine cycle. This parallel execution is shown in Figure 9.

Processing Unit Chip

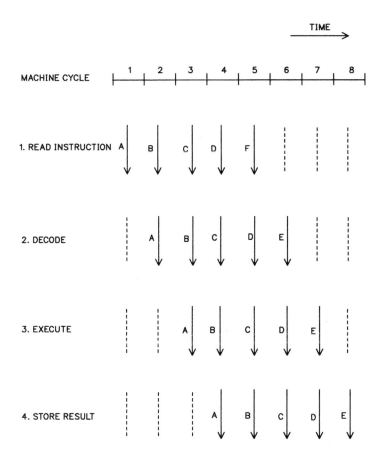

Figure 9. Pipelined instruction execution

The CPU chip performs pipelined instruction execution for either S/370, or for microinstructions. A mixture of both may be in execution within the pipeline at any time. Elements of the pipeline stages are:

- the addressing stage, where the address register addresses the next instruction to be executed. There are two address registers: the Control Store Address Register (CSAR) in micromode, and the Instruction Address Register (IAR) in S/370 mode. The latter is specified as part of the Program Status Word (PSW) in the S/370 architecture. The instruction thus obtained is loaded into an operation register (μOP1 Reg or /370 OP1 Reg).

- the source stage is controlled by the source decoders. Input to the source decoders are the μOP1 Reg and resp. the /370 OP1 Reg and the mode bits discussed in "2.2.3 Modes of Operation" on page 21. During this stage, the operands of the instruction are loaded into the ALU entry registers, the A- and B-Registers. There are several possible sources: DLS, Immediate Data Bus, Central Data Bus or Save Register (for A-Register only).

- the target stage is controlled by the target decoders. They are served by a duplicate of the operation registers, the μOP2 Reg resp. the /370 OP2 Reg. The operands are modified in the ALU or Shift Unit, and are loaded into the DLS-IN Register. The DLS-IN Register is set for all operations.

- The write stage uses the target decoders to load the ALU results from the DLS-IN register into the DLS array. The write address is copied from one of the source operands.

The source and target stage have separate mode bits to determine the mode of operation in each stage. In addition the source and target stage have their own sets of S/370 and micromode operation registers and cycle counters. There are 4 operation registers and 4 cycle counters in total.

The ALU/Shift Unit controls are target decodes from the logical point of view. For a minimum machine cycle and maximum processing speed the full cycle should be available for data manipulation. No time should be wasted for operation register decodes in the target cycle; the controls for the target cycle are decoded during the source cycle. The result is set into latches which directly control the gates in the target cycle. These control latches are logically equivalent to the Op2 Registers.

The microinstruction addresses are copied from the CSAR into CSAR Buffer 0 and from there into CSAR Buffer 1 during microcode execution in the "base level". The staggered CSAR buffers are a requirement for the repetition of a microinstruction after a successful TLB miss resolution. The TLB miss is detected in the second target cycle. At this time the address of the instruction causing the TLB miss is already propagated into CSAR Buffer 1.

ALU results are loaded into the DLS-IN registers at the end of the target cycle. They are written into the DLS array in the next cycle. Pipelining imposes no restriction to the microprogram; especially a result set into a DLS register in one instruction may be used in the next. This is achieved by a comparison of both DLS read addresses with the DLS write address. On match the result bytes to be written are selected as input to the A- and B-register from the Central Data Bus instead from the DLS. For this a special data path, the bypass, is provided.

Figure 10 shows three microinstructions and their execution in the different stages of the pipeline. Each microinstruction takes one cycle of the pipeline. The corresponding microcode assembler source code reads:

```
MV      W2,W5
SHIFT   W1,W4,amount
ADD     W2,W1
```

The CSAR addresses the μ-instruction in the Control Store. Normally, it is incremented by 1 after each cycle. The μOP1 register holds the μ-instruction in the source cycle to address the DLS, and the input selection to the A/B-registers. The A/B-registers hold two (or one) operands for the ALU operations. The μOP2 register holds the μ-instruction in the target cycle to control the ALU operation, the DLS write operation and the setting of the ALU Status latches; the DLS write control is also decoded from the μOP2 register. The DLS-IN register holds the ALU result for the write operation internal to the DLS.

Write and read of the same DLS register in one machine cycle is possible; the updated value is read.

2.2.5 Data Local Store Layout

The Data Local Store (DLS) is a 48 x 36 bits wide array. The 48 fullwords with 32 data bits plus 4 parity bits are grouped into:

- 16 General Registers, which are directly accessible by S/370 instructions.
- 4 Floating-Point Registers for S/370 Floating-Point Instructions. A Floating-Point Register is 64 bits wide, so 2 words are required for each register. These registers remain unused if the chipset configuration includes an FPU chip.
- 8 work registers for the base level. They are used as scratch space in microcoded E-phases.
- 8 work registers for the trap level. The calculations for the virtual to real address translation after a TLB miss can be performed without destruction of the intermediate values of an E-phase.
- 8 status registers, which contain copies of operand start addresses, a part of the Program Status Word, partial copies of Control Registers and the Micro Code Origin, MCO, address in the main store.

Figure 11 shows the complete layout. The 16 Control Registers required by the S/370 architecture are not part of the DLS, they are implemented in non-S/370 memory.

2.2.6 Micro Instructions

The Capitol chip set CPU chip employs microinstructions for interpretative execution of the less frequently used S/370 machine instructions.

Central Processing Units generally employ one of two microinstruction types: horizontal and vertical [Schu]. The bit positions in a horizontal microcode word control directly gates in the CPU dataflow; vertical microinstructions have similar characteristics as regular machine instructions. Often CPU's use both in a hierarchical fashion; in this case the horizontal microcode is referred to as nanocode or picocode. With the availability of Programmed Logic Arrays (PLA's) the distinction between horizontal microcode and hardwired logic starts to become fuzzy.

The Capitol chip set CPU employs vertical microcode. The microinstruction architecture is essentially unchanged compared to two predecessor machines:

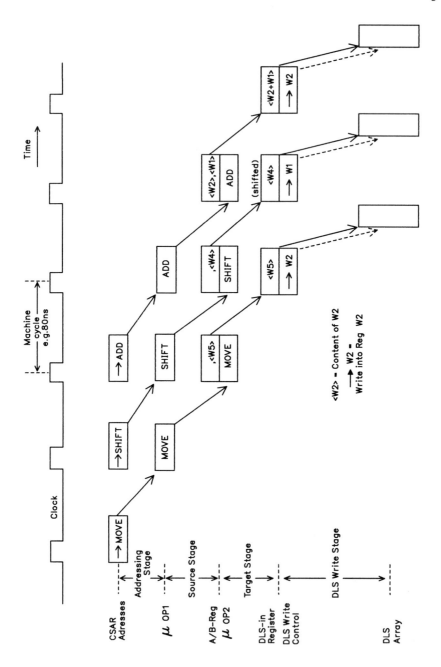

Figure 10. Pipelined μ-instruction execution

the IBM 4361, and the IBM 9370-90. In view of the expense to develop a new microinstruction architecture and the corresponding microcode, it is not uncommon for the design of a new CPU to utilize the microcode of a predecessor.

Most Capitol chip set microinstructions are tightly coded into 16 data bits (plus 2 parity bits) in length. Some branch instructions use a 32 bit long format. In the long format the second group of 16 bits may replace the content of the Control Store Address Register.

There are 2 reasons for selecting a 16 bit instruction format, the first being the cost of high speed microcode storage. In addition, I/O pins are a resource, which has both chip and package constraints. A microinstruction format of 32 bits (plus 4 parity bits) requires another 18 pins for the Control Store interface.

The number of pins for the Control Store interface will become irrelevant for future machines with a higher level of integration and an imbedded control store array. However, the number of bits per microinstruction will remain an important design consideration, this time trading control store array area versus the size of a more complex (larger) decoder, that is required for the compressed format. The decoding time for the more complex decoder (more stages of logic) is acceptable as long as it can be done one cycle in advance (see "2.2.4 Pipelining" on page 27).

The 16 bit instructions are divided into fields, that implement functions such as instruction type and operand address(es). The number of bits for each field, and the number of variations which can be achieved with each field, is rather small in comparison of the large number of functions required to interpret a complex architecture like S/370.

Compressing most microinstructions into a 16 bit format is achieved by including latches and registers outside of the micromode operation register (such as sign latches and bits of the S/370 operation register) into the function decoders, and by reducing the address capability to 8 DLS registers for many microinstruction types.

2.2.6.1 Data Local Store Addressing

Except for compression into a 16 bit format, the microinstruction set has many RISC characteristics. Especially, except for load and store microinstructions, operand addresses refer to the Data Local Store (DLS) or internal registers and latches. Also, most microinstructions execute within a single cycle.

Fitting the microinstructions into 16 bit results in a highly irregular, non-orthogonal microinstruction format. This applies in particular to Operand addresses, usually accessing the Data Local Store. The addressing of 48 Data registers requires 6 bits. The full addressing capability is available only for some sense and control, and for branch microinstruction types. ALU microinstructions, shift microinstructions, and load/store from/to main store microinstructions have a reduced addressing capability for 8 registers, requiring 3 bits per operand.

If the CPU chip executes in trap level, these 8 registers are the DLS registers T0..T7 (see Figure 11). In base level these 8 registers are either 6 direct and 2 indirect addressable registers or 8 direct addressable registers, depending on the S/370 Operation Code of the instruction being executed. The direct address-

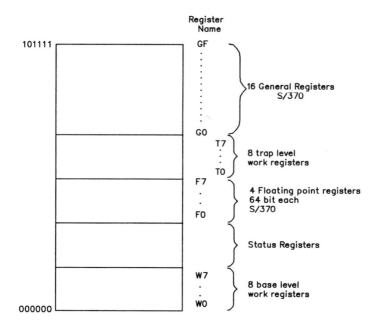

Figure 11. Data local store layout

able registers are W0..W7 resp. W0..W5. The indirect addressable registers are either the General or the Floating Point Registers just addressed by the contents of the R1-field of a S/370 instruction.

2.2.6.2 Microinstruction Types and Formats

Microinstructions belong to one of five basic types: data manipulation (such as ALU and shift instructions), DLS internal moves, setting and sensing hardware control registers and latches, main store accesses, and branching.

The ALU microinstructions permit the logical functions AND, OR and XOR and arithmetic functions for fixed-point, floating-point and decimal number representations. The ALU instructions are two-operand instructions, the first operand is also the target operand. Each operand address field is limited to 3 bits. The ALU functions are performed on fullword (32 bit) operands. Figure 12 shows the layout for the ALU instruction.

DLS internal MOVE microinstructions either write 4 Bytes, the low-order three Bytes or one Byte of the result register. Immediate operands of one Byte in length may be either written into the DLS, or into Control latches. Figure 13 shows the instruction layout for three of five types of MOVE instructions: MOVE of an immediate Byte into the DLS, MOVE with full address ability of the source operand (direct), and a MOVE with full addressability of the target operand (zone-indirect). Bits 10...12 are either used for the source or the target operand address.

Processing Unit Chip

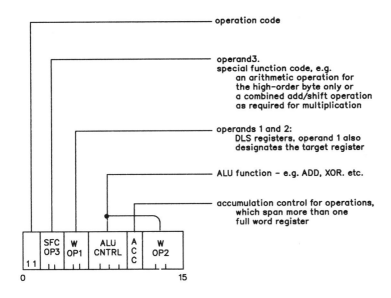

Figure 12. ALU microinstruction format

Several sense and control microinstruction types set and sense control latches and registers either on the CPU chip, or on other chips via communication over the Processor Bus. Control microinstructions set (and reset) CPU chip internal facilities from registers in the DLS. With the sense microinstruction CPU chip internal status information is sensed into DLS registers. All DLS registers can be accessed in this fashion. Other Sense and Control microinstructions set, sense and control the timer facilities on the CPU chip, perform control store read and write operations, or Processor Bus operations.

Sense and Control microinstructions manipulating data or facilities on other chips are executed by putting an 8 bit "command" onto the Processor Bus and activating the control line CMD_VALID (see "2.10.1 Processor Bus Connections" on page 128). A command occupies bit positions 0..7 on the 32 bit data bus. Frequently, an address is transmitted within the same machine cycle in bit positions 8..31. An example for a command operation over the Processor Bus is shown in Figure 88.

Processor Bus commands may be issued by either the CPU chip (via microinstructions) or alternatively by the BBA chip.

Accesses to the cache take two cycles on the Processor Bus, if the data do not cross a doubleword boundary and the data are in the cache. In case the data are not in the cache, they will be loaded into the cache by a cache line load operation. The line load may require a cast-out operation, which writes a cache line of 64 Bytes into main store, to gain space for the line requested. The number of Bytes for a virtual access varies from 1 to 4 Bytes.

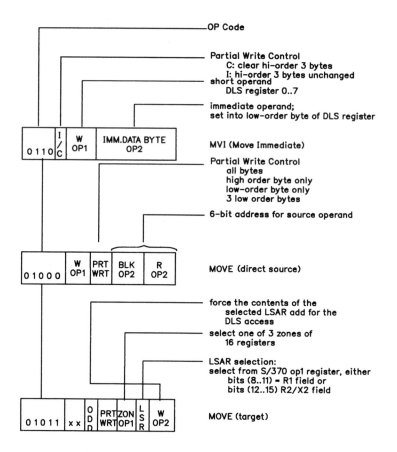

Figure 13. MOVE instruction layout

2.2.6.3 Reducing the Number of Branch Microinstructions

An ALU microinstruction requires one machine cycle, while a successful branch needs 3 machine cycles. Thus more time, or at least as much time, may be spent for branching as for the ALU functions itself in a short micromode loop. (Almost the same ratio may be found in the execution of S/370 storage-to-storage instructions, where the fetch and store take 2 cycles each and the branch, controlling the loop, takes another three cycles).

For elimination of these branches two facilities have been incorporated:

- the Force On Boundary facility in ALU instructions and virtual cache accesses to control a loop of two or more microinstructions, and
- the Suppress Set Op-register for repetition of single ALU instructions.

The Force On Boundary provides a simple, but useful branch facility: the three low-order bits of the Control Store Address Register are forced to zero, depending on a loop end condition. This end condition may be the length code field of a S/370 instruction being executed. It is decremented with each cache access. The forcing of the address does not become effective for the next instruction due to the pipelined operation, but for the second next instruction. With the Force On Boundary facility a loop of up to 7 microinstructions can be controlled. An example is given in Figure 14. It shows a part of the microcode for the E-phase of the S/370 storage-to-storage instruction Move Character (MVC). Starting on a 8-instruction boundary, the sequence of FLL, STLL instructions is executed (at least once), till the S/370 length code in the LSAR01, which is identical to the S/370 OP1 register bits (8..15), is exhausted.

```
FLL    W3, SAR2, +4, FOBL01_* FETCH 1..4B FROM OP2
                              EFL = MAX 4, DEPENDING ON LSAR01
                              INCREMENT SAR2 BY EFL
                              NO LSAR01 DECREMENT
                              FOB TO '0000' IF <LSAR01> >= EFL
STLL   W3, SAR1, +4.D         STORE 1..4B TO OP1
                              EFL = MAX 4, DEPENDING ON LSAR01
                              INCREMENT SAR1 BY EFL
                              DECREMENT LSAR01 BY 4
                              NEXT I-PHASE, NO CC CHANGE
CTRLI  NEXT, B'00001000'      RETURN WITHOUT CC-SETTING
```

Figure 14. Example for a FOB loop

The Suppress Set Op-register feature keeps the contents of the micro opcode register, and of the Control Store Address Register unchanged. This feature is activated for multiplication of operands of fullword length or less; multiplication is implemented as repetitive addition and shift. The addition and shift is performed by a single microinstruction.

The Force On Boundary and the Suppress Set Op-register are activated by Special Function Codes.

2.2.7 Data Flow Logic

2.2.7.1 ALU, Shift Unit and DLS

The A- and B-register can be viewed as DLS data-out registers, but may be loaded alternatively from the Immediate Data Bus, the Central Data Bus, or the Save register (A-register only). For two-operand microinstructions, the first operand, which is also the target operand, is put into the A-register and the second operand is set into the B-register.

The Central Data Bus may be input for 3 different sources: data, fetched from the MMU, participating in arithmetic or logical operations, data sensed out of hardware registers (e. g. STAR's), or as bypass for data to be written into the DLS and simultaneously selected as an operand in the next cycle.

The Immediate Data Bus is a collection of many sources, e.g. immediate Bytes out of S/370 or micro instructions, data sensed out of hardware registers, the displacement field as specified in S/370 instructions with storage operand(s), or CSAR data, e. g. link information.

The operands are loaded into the A- and B-register in the source stage cycle. In the target stage cycle they are manipulated in (or passed unchanged through) the ALU, Shift Unit, or modify-bit-0 block. The output of one of these units may be selected as input to the Central Data Bus; the A- and B-register itself may be selected as input for the ALU output multiplexer. The B-register is gated to the Central Data Bus in the MOVE microinstruction; the A-register will be selected conditionally in the divide operations.

The ALU performs the basic functions OR, AND, XOR, and ADD. With the inverted output of the B-register and a forced carry into the lowest-order ALU bit position, the SUBTRACT function is achieved.

Decimal data are kept as packed decimal data in storage and in registers. The decimal values 0..9 correspond to the hexadecimal values X'0..9'; the sign digit is replaced by a zero digit during the calculation. For decimal additions and subtractions the "Excess 6" method is used. "6" is added to each digit of the B-register operand, which will never generate a carry for valid decimal data; "6" is subtracted from each digit of the ALU result, unless it did pass a carry to the next digit.

The various inputs forcing zeroes and ones at the ALU entry are mostly activated bytewise, triggered by microinstructions with special function codes. The 32-bit wide ALU can also be used for 24-, 16- or 8-bit operations.

The Shift Unit may receive input for shift controls from the ALU for the S/370 logical and arithmetic shift instructions (Shift Left Logical), and through a Control microinstruction that sets a shift amount from a DLS register. Shift unit controls may be set

- from the ALU Status dependent on the number of leading zero digits in floating point numbers,
- from the Immediate Bus for microinstructions, specifying the shift amount directly,
- from the Immediate Bus for the S/370 mask instructions, such as Insert Characters under Mask (ICM), where the M3-field of the instruction is encoded into shift amounts, one for each Byte.

Basically, two registers control the operations within the Shift Unit:

- the Shift Amount (SA) register, which controls the data gates of the shift unit itself.

- the Characteristic Difference (CD) register, whose main purpose is the support of floating point operations (for configurations without an FPU chip). The contents of the CD register is a "raw" shift amount, which may be a "complemented" shift amount. It is unpredictable, whether the floating point number with the smaller characteristic will be subtracted from the one with the larger characteristic or vice versa. In case the smaller characteristic is subtracted from the larger one, the shift amount is "true", otherwise a "complement" shift amount is decoded from the CD-register.

The block diagram in Figure 6 shows, that some control pattern for the shift unit are set directly into the SA register; other control pattern are set into the CD register, requiring further decodes. The shift amount either specifies the number of digits to be shifted (as it is required for floating point numbers or packed decimal data) or the number of bits to shifted.

Data can be shifted left or right; the shift operations set the ALU zero/nonzero status latch. Bits and digits shifted out are not retained; the vacant positions are filled with zeroes.

2.2.7.2 Bus Unit and Processor Bus Operations

The bus unit (see Figure 6) is a part of the CPU chip that provides the connections to similar bus units on the MMU, FPU, and BBA chips via the Processor Bus (see "2.10.1 Processor Bus Connections" on page 128). Except for the page-look-ahead for a complete operand, the bus unit does not provide circuitry for data crossing boundaries, e.g. doubleword-, cache line-, or page-boundary. The bus unit always issues a single request to the MMU; for data to and from the Cache the MMU will merge data from both sides of the boundary. (Internally the MMU performs 2 cache accesses and merges the data. The CPU has to wait one or two extra cycles). For accesses to main storage, which are passed unchanged through the MMU, the data have to be aligned. This is ensured by microcode.

There are three different types of storage accesses to the MMU:

- virtual accesses, which are the standard access for S/370. The address is translated in the TLB on the MMU chip ("2.5.2.1 Overview" on page 81) from a virtual to a real address. The real address out of the TLB addresses the cache array on the MMU chip.

- real accesses, which bypass the TLB. This command is issued as access to fixed main memory locations, e.g. for the old and new PSW in a PSW swap. The address is directly applied to the cache.

- accesses to the main store, bypassing TLB and cache. For these commands the MMU performs only a bypass function. They are used for array initialization and for accesses to non-S/370 objects.

The addresses which are set into IAR, STAR1 or STAR2 from the central data bus or modifier may be set in parallel into SAR and BSR. In this way an address set in one cycle, may be put on the Processor Bus in the next cycle.

The Processor Bus is used for command/address transfer (in the first cycle) and data transfer (in the second or later cycle). In a similar way the key-status bus serves two purposes: in the first cycle the key is sent together with the command and in the data cycle a status is received.

The Status bus is used to signal indications for TLB miss, protection/ addressing violation, any malfunction in the unit, which received the command, and S/370 address compare match. The latter results in an exception at the end of the S/370 instructions, all other indications cause a trap immediately.

The bus unit contains the address registers for the S/370 instruction and two operands: IAR, STAR1 and STAR2. The Storage Address Register, SAR, and the Bus Send Register, BSR, keep the address for the current access. Two registers have been implemented for the same function to have the minimum delay from the register to the processor bus. Moreover, Byte 0 of the BSR receives the command for the processor bus operation. The modifier, MOD, increments or decrements the address by the number of Bytes transferred. The update is suppressed in case of a nonzero status from the MMU; so the original address is available for the repetition of the access. The modifier is also able to increment an address by 4 KByte to perform the Page-Look-Ahead access as described in "2.2.3.3 Forced Operations" on page 26.

Each of the four modes described in section 2.2.3 may perform functions, which involve the bus unit.

In the S/370 mode the bus unit encodes the bus commands from the S/370 operation register and the S/370 cycle counter for virtual address main store access. The number of Bytes is either fixed length or depends on S/370 operation register bits(8..15). The address is taken from the IAR, STAR1 or STAR2.

In the micro mode the bus unit supports several types of micro instructions, which activate the Processor Bus for fetch and store operations and for senses and controls to other units on the Processor Bus. The controls and senses to other bus units including the MMU initialize and sense status from other units and start I/O commands.

The bus unit is also involved in the command transfer, which sends the contents of the S/370 operation registers bits (0..15) to the FPU chip during a floating-point instruction as described in "2.6.3 FPU Interface and Communication" on page 100. The S/370 instruction bits(0..15) are put in the right part of the word transferred, the left part contains the bus command and two digits, which specify the sending and receiving unit address respectively.

Bus unit decoders may trigger forced operations such as traps or exceptions in case of a nonzero status from the MMU and it may force a Page-Look-Ahead access in conjunction with an initial access in E-phases in micromode and S/370 mode.

The bus unit supports the forced operation "Transfer", which reads microinstructions from the non-S/370 memory or the EPROM address space. Which medium is accessed depends on a latch that is set and reset by micro instructions; after power-on this latch is initialized to access the EPROM address space.

2.2.8 Prefetch Buffer

Instruction prefetching is a mechanism which utilizes the sequential nature of the instruction stream to reduce the access time to the main store or cache. In the Capitol chip set the cache access time takes 2 machine cycles. To shorten the time required to get the next sequential instruction from the cache, the CPU chip incorporates a Prefetch buffer that is automatically loaded with instructions following the one just being executed, see Figure 15.

Prefetching instructions does not work when the sequential instruction stream is broken, e.g. in the case of branches. Here a flushing and reloading of the prefetch buffer causes some overhead. To reduce this, the prefetch buffer has to be kept small; the size chosen for the Capitol chip set is 2 words, or 8 Bytes. The Prefetch buffer hit rate is rather high; the performance degradation due to prefetch buffer misses is only a few percent for typical commercial applications and less for engineering scientific applications. Increasing the prefetch buffer size will decrease the performance.

The prefetch buffer may be reloaded word by word, as the instructions are taken into the S/370 operation register, or may be filled completely in two consecutive accesses. Instruction fetches always specify a word boundary. There is a fixed address assignment for the left and the right buffer. Instruction words fetched on an even address are set into the left buffer, the others are set into the right buffer; all accesses are on a doubleword boundary. A successful branch loads the Prefetch-buffer completely; these accesses are part of the branch operation.

A coarse S/370 instruction address register (IAR) with the two low order bits set to zero reloads the prefetch buffer. The low order bits select the instruction halfwords out of the prefetch buffer at execution time. They move the first halfword of each S/370 instruction from the Prefetch buffer into the S/370 Operation Register and gate the second and third halfword (if any) to the ALU for the base/displacement calculation.

Whenever a micro instruction specifies switch to S/370 mode after initialization or PSW swap, the CPU chip first enters the S/370 FOP mode for a complete Prefetch buffer load. After completion of the load operation, the CPU-chip enters the S/370 mode. As a rule, instruction execution includes reloading the Prefetch-buffer, so the next instruction is always completely available at the beginning of its instruction phase (I-phase). S/370 instructions in RX, RS, RRE, S and SI format, which are 4 Bytes long, cause a word access to reload the Prefetch-buffer. RR format instructions are 2 Bytes long. If an RR instruction starts at a word boundary, it does not cause a reload. If it starts on an odd halfword address, a reload takes place, so the word of the Prefetch-buffer, from which the RR format instruction was taken, is filled again. Similarly SS

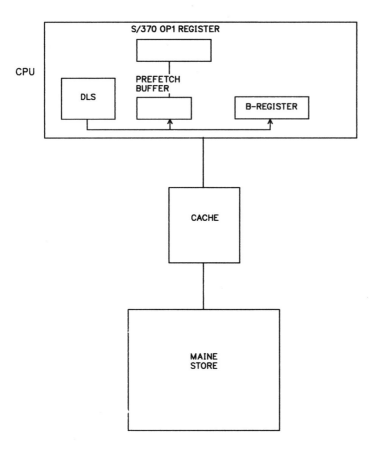

Figure 15. Prefetch buffer

and SSE format instructions that are 6 Bytes long will trigger two reload accesses when starting on an odd halfword address, and one reload when starting on a word boundary.

One cycle of the Prefetch-buffer reload overlaps with the address calculation for non-RR format instructions; in RX instructions with a nonzero index register two cycles overlap with the address calculation. The same is true for SS and SSE instructions with one Prefetch buffer reload. For S/370 instructions with a microcoded E-phase the transition to micromode takes place after the Prefetch-buffer reload(s). Thus, there is no mode transition during an access to the MMU chip, which eases repetition of an access after a TLB miss.

Exceptional conditions such as TLB miss, protection or addressing violation occurring during a Prefetch-buffer reload require special attention. They are detected ahead of the instruction execution of an instruction from that location. At the time of the detection of the exceptional condition it is unknown, whether the unavailable word is really part of an instruction to be executed. It may happen, that the instruction, which causes the Prefetch buffer reload with exception, or even the following instruction, is a successful branch with the branch target in an available page. The exception has to be ignored in this situation.

Therefore a two-level mechanism is implemented for TLB, protection and addressing violation. In the first level, a latch is set upon detection of each of the exceptions. These latches are reset, if a successful branch or a program interrupt, e.g. fixed-point overflow, occurs before the S/370 low-order instruction address advances to the Prefetch-buffer not loaded.

Whenever the instruction address advances to the part of the Prefetch buffer, which has not been loaded, one out of a second set of latches is set. Activation of one of these latches causes the proper S/370 program interrupt.

Even so instruction modification is not a good programming practice, the S/370 architecture does not exclude it. Therefore a store operation into one of the instructions just prefetched causes invalidation of the complete Prefetch-buffer. Micromode gets controls and causes a complete Prefetch-buffer load with the updated instruction.

2.2.9 Floating-Point Coprocessor Interface

For fast floating point instruction execution, the Capitol chip set includes an (optional) floating point coprocessor chip, the FPU chip. The chip set will work without the FPU chip, and can execute floating point instructions through microcode, although at a lower speed. Working with the FPU chip, the CPU transmits floating point instructions to the FPU for execution. The Processor Bus is the main communication path between the CPU and the FPU. Special Processor Bus commands for floating point register initialisation and sensing, and for floating point instruction execution interface the FPU chip to the CPU chip.

Besides the Processor Bus, on which the instruction is sent and data are exchanged, some additional control lines

- send the S/370 condition code to the CPU chip,
- permit the three party communication (see "2.6.3 FPU Interface and Communication" on page 100),
- indicate 'busy' and exceptional conditions to the CPU chip.

These extra lines are parity checked.

The S/370 architecture defines 4 unique floating point registers, each 64 bit long. They are in addition to the 16 General Registers. In a configuration with the FPU chip the floating-point registers are located on that chip. The Data Local Store (DLS) on the CPU chip contains a duplicate of these registers that are used only in an FPU-less configuration. Most of the S/370 floating-point instructions set the S/370 condition code. The condition code latches are implemented only once, on the CPU chip. For each floating-point instruction, which sets the condition code, the FPU sends the 2 condition code bits together with a set pulse over the Processor Bus to the CPU.

For a floating-point instruction in RR format it is sufficient to send a message command from the CPU chip to the FPU chip. This message command complies with the Processor Bus standard. The first Byte contains the command, the second byte specifies the sending and receiving bus units (one digit each), and the next two bytes contain the S/370 instruction.

2.3 Timer Support

Wolfgang Kumpf

2.3.1 Introduction to the Timer Functions

The CPU chip contains the System /370 timer hardware support functions. They, in combination with microcode, implement four possibilities to measure time: the time-of-day clock, the clock comparator, the CPU timer and the Interval timer. Resolution of the TOD clock and CPU timer is 2 μs; it is 1/300 sec for the interval timer.

The time-of-day clock provides a high-resolution measure of real time suitable for the indication of date and time of day. The cycle of the clock is about 143 years. The operation of this clock is not affected by any normal activity or event in the system.

The clock comparator provides a means of causing an external interruption (to be handled by an operating system interrupt routine) when the time-of-day clock exceeds a clock comparator value specified by the programmer. Both values are considered as unsigned binary integers and the comparison follows the appropriate arithmetic rules. The contents of the clock comparator are initialized to zero by initial CPU reset.

The CPU timer provides a means for measuring elapsed CPU time and for causing an interruption when a pre-specified amount of time has elapsed. In the basic form, the CPU timer is decremented by subtracting a one in a model dependent bit position. Whenever the CPU timer value gets negative (bit 0 of the CPU timer is one), an external interruption is requested by the CPU timer. When both the CPU timer and the time-of- day clock are running, the stepping rates are synchronized such that both are stepped at the same rate. The CPU timer is set to zero by initial CPU reset.

The interval timer is a binary counter that occupies a word of real storage at location X'000050' and is treated as a 32-bit signed binary integer. In the basic form, the contents of the interval timer are reduced by one in bit position 23 every 1/300 of a second (which corresponds to the power net frequency of 50

Hz and 60 Hz). A request for an interval timer interruption is generated whenever the interval timer value is decremented from a positive or zero number to a negative number. The interval timer is not necessarily synchronized with the time-of-day clock. The interval timer contents are updated with the appropriate frequency whenever other machine activity permits.

2.3.2 Format of the Timer Binary Counters

The time-of-day, the clock comparator and the CPU timer are binary counters with a basic format shown in Figure 16. The bit positions of these clocks are numbered 0 to 63, corresponding to the bit positions of a 64-bit unsigned binary integer.

In the basic form, the time-of-day clock is incremented by adding a one in bit position 51 every microsecond. In the CPU chip this clock is implemented as a counter with bit positions 0 to 50 and is incremented by adding a one every 2 microseconds in bit position 50 (see part b of Figure 16.

The clock comparator has the same format as the time-of-day clock and in the basic form it consists of bits 0 to 47, which are compared with the corresponding bits of time-of-day clock. On the CPU chip a counter is implemented in the same way as for the time-of-day clock and runs under the same conditions. This counter is loaded by micro-code with the complement of the comparator value. As the counter is incremented by one in bitposition 50 every 2 microseconds an overflow occurs by reaching the comparator value. A comparator request is set and stays active until a new counter value is loaded. This comparator request is handled by micro-code and may cause an external interruption with interruption code X'1004'.

The CPU timer also is a binary counter with the time-of-day format, except that bit 0 is considered a sign. In the basic form the CPU timer is decremented by subtracting a one in bit position 51 every microsecond. On the CPU chip it is implemented as a counter with bit positions 0 to 50 and is updated by adding a one in bitposition 50 every 2 microseconds. As this counter also is incremented, the the complement of the elapsed CPU time to be measured is loaded into that counter by micro-code. If the sign bit gets positive (bit 0 of the counter is zero) the CPU timer requests an external interruption with interruption code X'1005' as long as the sign bit stays positive.

To activate the update of the interval timer at storage location 80 a 13-bit binary counter is implemented (see Figure 16). This counter also is incremented by adding a one in bit position 12 every 2 microseconds and raises a 'Location 80 Update' request for one cycle every 1/300 seconds (which is realized all 3.332, 6.666 and 10.000 milliseconds).

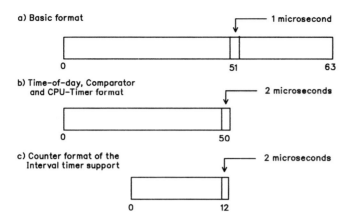

Figure 16. Format of the timer binary counters

2.3.3 Functional Description and Block Diagram

The major components of the timer hardware support logic are the input multiplexer, the four independent registers for the counter values of the timing facilities described above, the selector logic, the modifier logic, the control logic and the parity check logic. Figure 17 shows the data flow block diagram. The figures shown next to the connections indicate the number of data and parity bits.

The control logic section has 2 functions. It controls the timer internal dataflow and decodes and reacts to timer related CPU commands (control microinstructions) for data exchange via the Central Data Bus. The control logic signals are generated by an interval cycle counter (stepping synchronously with the chip set clock cycle) and by decoder logic for the CPU commands.

The internal cycle counter allocates timeslots for each of the four registers containing the timer facility values (see allocation scheme in Figure 18. The register content of an addressed facility is switched through the selector logic and updated in the modifier circuitry by adding a binary one in bit position 50. The updated value is written back into the according register with the next clock cycle.

In this way each of the register contents is updated with a repetition rate of 2 microseconds. The exact repetition rate is achieved by this fixed allocation of time slots to the different timing facilities.

If a timer related control microinstruction command has been decoded by the control logic it will be executed during the time slots which are not engaged by the update of the counter values. All external operations have to be delayed as long as internal updates are performed to avoid invalid data exchange.

The timer hardware support logic is completely checked by adding one parity bit to each data byte. Input data from the Central Data Bus are checked for

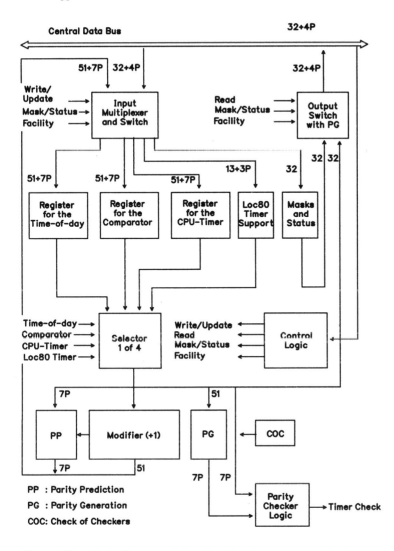

Figure 17. Block diagram of the timer hardware support logic

correct parity and the appropriate parity bits for output data to the Central Data Bus are generated. During the internal updates the parity bits of the counter values stored in the registers are modified by the parity prediction logic and, in parallel, checked by the parity checker logic.

Parity errors may activate two kinds of error indications, the Timing Facility check and the Interval Timer check. This will raise an exception condition that generates a Machine-Check Interrupt.

The parity checking circuitry may be tested bytewise by stimulating the "checking-of-checkers" feature (see COC logic block in Figure 17). This is done by a special test routine which is part of the IML test programs.

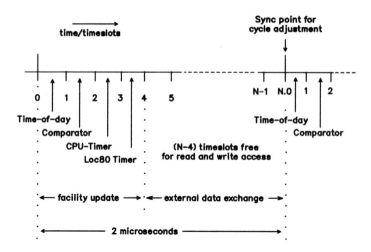

Figure 18. Timeslot allocation scheme

2.3.4 Communication with the CPU

There are 3 types of communication between the timer hardware and the remaining CPU chip functions: a) write the binary counter values, mask bits or status bits, b) read these data and c) activate dedicated request and exception signal lines. The write and read actions are performed by special "Control" and "Sense" microinstructions.

Specific types of "Control" microinstructions set the counter registers to the proper start values. In a similar way a timer internal "Stopped state" and the timer mask bits may be set and the "Location 80 update" request (interval timer update) may be enabled. During the S/370 architectured "Stopped state" the binary counters representing the CPU timer and the interval timer are not incremented. With the timer mask information the comparator request and the CPU timer request will lead to an external interruption condition or not.

All information set by "Control" micro instructions can be read back by corresponding "Sense" micro instructions. Since the CPU-chip data flow is 32-bit wide, the 'Control' and 'Sense' micro instructions to write and read the timing facility registers requires two internal time slots (two machine cycles) for data transmission.

2.3.5 Programmable Clock Cycle Time

To accommodate semiconductor technology advances resulting in improved cycle times, a wide range of different clock cycles is possible with the Capitol chip set.

For this an additional mechanism in the timer hardware has been implemented that adjusts the timer control logic to the various clock cycle time to perform the update exactly in two micro-seconds and to guarantee sufficient accuracy of the timer functions (notice the adjustment point in Figure 18. Therefore a time deviation of the timer functions only depends on the tolerance of the oscillator generating the timer clock pulses.

To perform this adjustment specific bit patterns are defined to represent the possible clock cycle times. The timer hardware may be programmed by loading the pattern according to the chosen cycle time with a special micro instruction during IML whenever the default clock cycle of 80ns is not the desired one.

2.4 Memory Management Unit Chip

Horst Fuhrmann

2.4.1 Overview

In entry systems like the IBM PC, the CPU chip addresses main store directly; instruction and data addresses in the CPU are identical to those in main store.

More sophisticated systems invariably have facilities for Dynamic Address Translation (DAT) to implement virtual memory. Address translation adds significant complexity, and requires a disk for external page storage, but provides superior facilities for the operating system and the end user to manage real main storage. Usually a CPU has the facility to switch address translation on or off, and thus to work in either virtual or real addressing mode.

As CPUs become faster, there is an increasing gap between machine cycle time and main store access time. This gap becomes larger, as main store sizes grow. Implementations in dynamic RAM technology require additional bus drivers as well as error correction and redundancy control circuitry. To prevent signif icant performance degradation due to waits required for main store access, the use of a small, high speed main store buffer, the Cache, has become common. Again, the necessary cache control logic adds significant complexity to the design.

Finally, the S/370 architecture provides a high level of data integrity by implementing real main store protection via storage keys. The storage keys are kept

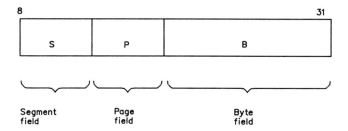

Figure 19. S/370 virtual address format

in a high speed buffer, the key store, that is accessed in parallel with each main store access.

The Capitol chip set implements the hierarchy of storage devices (virtual address translation, cache, key store) by combining their control circuitry into a single, highly integrated design. This circuitry constitutes the major part of the logic on the MMU chip. Additional parts of the storage hierarchy, e. g. external page store or the file system, are managed by the operating system, and require only few facilities in the Capitol chip set.

2.4.2 Storage Hierarchy Elements

2.4.2.1 Virtual Storage Addressing

Main store addresses generated by the CPU are either instruction addresses, or operand addresses resulting from an address calculation process. In a system with virtual address translation, both instruction addresses and effective operand addresses are virtual addresses that need to be translated into real addresses before accessing main store. A virtual address consists of a number of fields; for the /370 architecture the virtual address format is shown in Figure 19.

The translation of a virtual address into a real address occurs with the help of 2 table lookup operations. For this purpose the supervisor maintains 2 types of tables in real main store. The starting address of the first one, the segment table, is contained in a CPU control register (segment table origin). The segment field in Figure 19 contains an offset to address an entry in the segment table. This entry contains a pointer to one of many page tables. The page field in Figure 19 contains an offset to address an entry in the page table. This entry contains the upper 12 bit of the desired real address; it is concatenated with the 12 lower bits of the virtual address to form the real main store address.

This segment/page table translation scheme is shown in Figure 20. Obviously, 3 accesses to main store are required to retrieve a single item (instruction or operand). Some architectures, e. g. the VAX architecture, avoid segment tables and use a single (large) page table instead. However, because of the size of the resulting page table, at least for problem state programs the page table has to be kept in virtual storage instead of real main storage, resulting in the same number of main store accesses as with a segment/page table approach. Another alternative is the "inverted page table" employed in the IBM 6150 architecture. Here again a single page table is used. Different from the VAX architecture, the page table is kept reasonably small by storing entries only for those page frames that reside in real storage. However, due to a synonym problem and a limited search requirement, the number of main store accesses is still about the same as with the segment/page table approach employed by the S/370 architecture. It is due to these difficulties that some recently introduces new architectures like the Intel 80386, the HP Precision, and the Motorola 88000 utilize segment/page table translate mechanism that are quite similar to the S/370 mechanism.

2.4.2.2 Translation-Lookaside Buffer

To avoid the overhead of 3 main store accesses for each real access, a "Translation-Lookaside Buffer" (TLB) is used. This is a high speed array that stores the most frequently utilized pairs of virtual and real addresses. It is addressed in parallel with each main store access. Assuming it contains the desired address pair, a single access is sufficient to store/retrieve an item in main store.

The TLB may take the form of a small associative memory, usually implemented as a set of registers that are read out in parallel, see Figure 21. The virtual part of each address pair is compared in parallel in all registers against the actual virtual address, and, if a match is found, the corresponding real address of the address pair is utilized. The registers usually contain a few control bits in addition to the address pair to implement a LRU (least recently used) or other replacement algorithm in case of a TLB miss.

Associative memories are expensive to implement, and therefore frequently replaced by alternative solutions. Thus Translation-Lookaside Buffers are often implemented as a "set associative" TLB that uses several small regular random access read/write arrays (memories).

For this assume a TLB design as shown in Figure 22. Imagine a 16 MByte virtual address space subdivided into 128 subspaces of 128 KByte each, as shown in Figure 23. Each subspace thus contains 32 pages of 4 KByte each. The 32 page locations in each subspace are characterised by a number n, where $0 \leq n \leq 31$. Now assume that at a given moment less than 32 pages are in use (the working set is smaller than 32 pages), and that each of the pages happens to have a different value of n. In this case the TLB shown in Figure 22 is able to hold all virtual addresses of the working set. During a particular access the TLB entry with the address n is read out. The left field of the entry should have the same value as the high order bits (bit 8...14) of the virtual address. The right field of the entry contains the real address.

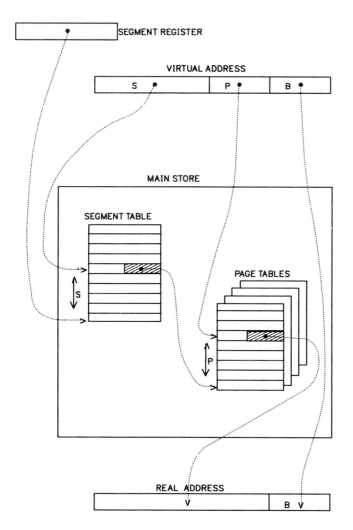

Figure 20. Segment/page table

In reality, the value n may occur more than once among the 32 (or less) pages of the working set, resp. in the 128 subspaces of Figure 23. Therefore the small array is duplicated as shown in Figure 24. The TLB will work as long as there are no more than 2 virtual addresses with the same value of n among the pages in the working set. With 4 parallel arrays n may occur up to 4 times. Practical experience indicates that 2 parallel arrays guarantee already a sufficiently small TLB miss rate.

Memory Management Unit Chip

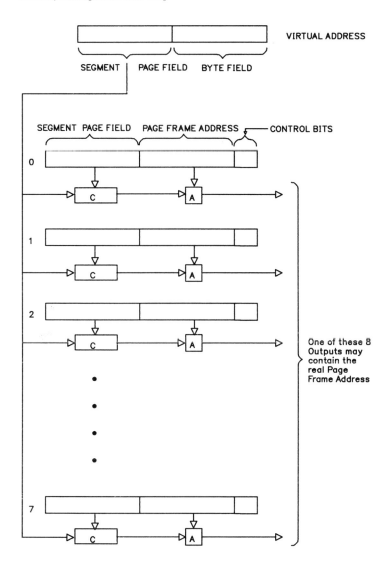

Figure 21. Full associative translation-lookaside buffer

2.4.2.3 Cache Operation

To compensate for the mismatch in CPU machine cycle time and main store access time, frequently used instructions and operands are kept in a high speed cache. The cache is subdivided into blocks of identical size called "cache lines", typically 64 bytes in size (see Figure 25). If a desired item is not found in cache, a complete cache line is loaded from main store into cache. Since the cache is probably already filled, one of its cache lines has to be moved into main store to free up space.

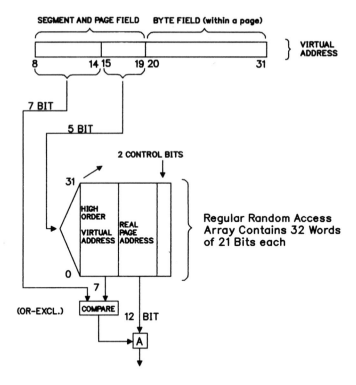

Figure 22. Non-associative translation-lookaside buffer

A S/370 cache works with real addresses. A "cache directory" maintains the addresses of all cache lines presently stored in cache. The cache directory solves a similar problem as the TLB, except that each entry contains not a real address but a cache line address of the cache array. A 4-way set associative implementation of the cache directory is common. For this, the cache array itself is split into 4 equal parts, called cache compartments. Each compartment is addressed by 1 of the 4 arrays in the 4-way set associative cache directory. This design is shown in Figure 26.

2.4.2.4 Key Store

Storage protection, the prevention of accidental or illegal access to main store outside the address range assigned to a particular process, is an important data integrity and data security feature. Most architectures provide storage protection only as part of the virtual memory address translation mechanism, usually as part of the segment and page tables. The S/370 architecture is unique in that it provides storage protection to real main storage as well.

For this, real main storage is divided into 4 KByte blocks (page frames). Each block is assigned a 4 bit storage protection key. A separate high speed buffer, the "key store" contains one entry for each 4 KByte Block in main store. Each

Memory Management Unit Chip

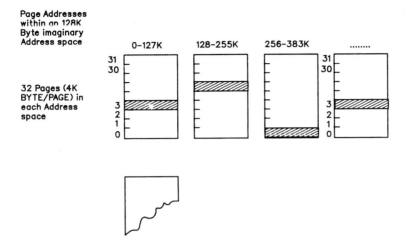

Figure 23. Imaginary subdivision of a 16 M byte address space into 128 imaginary address spaces of 128 K byte each

entry contains a 4 bit key (plus some additional bits). The current Program Status Word (PSW) in the CPU contains a 4 bit storage protection key field. During each access to main store the 4 bit storage protection key field in the PSW in compared with the entry in the key store that belongs to the accessed 4 KByte main store block. A mismatch generates protection exception. The key store is shown in Figure 27. Each entry has 3 bits in addition to the 4 bit storage key.

2.4.2.5 Combined Operation

The TLB uses virtual addresses as input. The cache directory and cache require real addresses. It may thus be assumed that the access to the TLB, the cache directory, and the cache itself have to occur in series (see Figure 28).

The partition of the cache into 4 compartments in Figure 26 permits a parallel access of cache and cache directory. This is important since the resulting cache access time significantly influences the total CPU performance. It is possible to perform the access to the TLB in parallel as well. Such a design is shown in Figure 29.

The right part of the virtual address, the byte field is identical in both the virtual and real address. It is therefore feasible to address the cache directory and also the individual compartments of the cache itself while the result of the address translation with the help of the TLB is still outstanding. Thus the parallel accesses shown in Figure 30 are feasible. Only the compare operation, and the key store access have to occur subsequently (the same is true for setting the LRU bits in the cache directory). A further improvement is possible if we add a field to the TLB that duplicates the storage protection key.

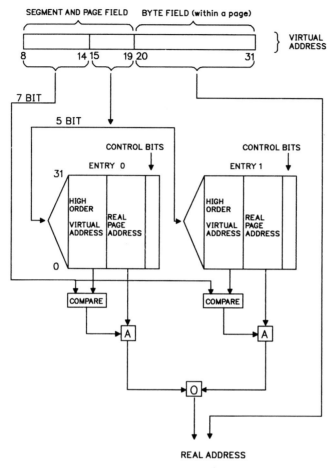

Figure 24. 2-way "set associative" translation-lookaside buffer

2.4.3 MMU Chip Data Flow

The TLB, cache, cache directory, and key store functions described in the preceding section are implemented on the MMU chip. The data flow shown in Figure 31 can be divided into three major parts. These are the command section (in the left half), the translate table and directory section (at the top), and the storage section (at the lower right half). The storage section consists of the 4 KByte key storage (for 4,096 keys each covering a 4 KByte area in real main store), and the 8 Kbyte cache storage with its associated aligners (shift units).

The MMU chip communicates with the CPU Chip and other Processor Bus units (such as BBA or the FPU chips) via the bi-directional Processor Bus (see "2.10.1 Processor Bus Connections" on page 128) shown at the bottom of the

Memory Management Unit Chip

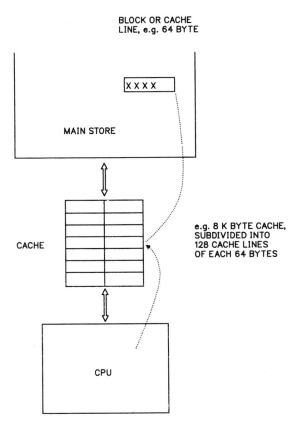

Figure 25. Cache concept

data flow. Whenever the control line "CMD_Valid" (see Figure 86) is active, the MMU accepts Processor Bus bits 0...7 into its command register, and Processor Bus bits 8...31 into its address register. Command and address are required for any storage access.

The same information is routed in parallel to the STC interface register on the MMU chip. It places this data through the STC multiplexer onto the STC bus that interconnect the MMU and STC chips (see top of the data flow). This parallel feeding prepares for the eventuality that the addressed location is currently not available in cache storage. If it is available, the STC chip access is not started. For fetch operations, the STC-interface register latches up the data received from the STC chip. This data can be routed to the cache storage or directly to the Processor Bus.

Store Operations require 2 Processor Bus cycles. The data to be stored are received from the Processor Bus in the cycle following the command and address transfer, and are placed into the data register. The command in the command register stimulates the decoder which drives the operation control, and activates all gates in the various data paths.

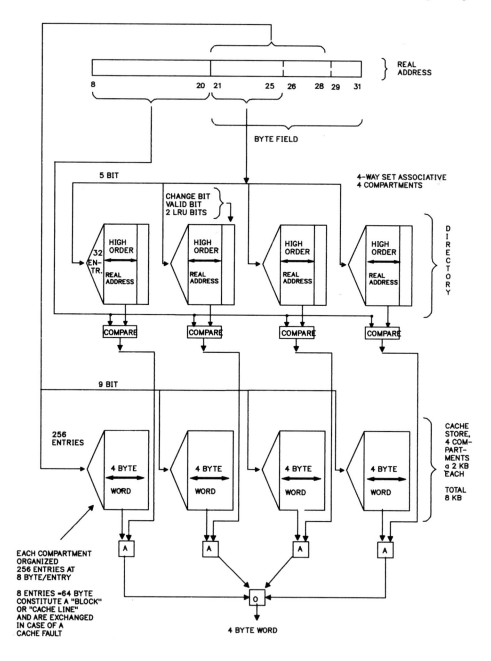

Figure 26. S/370 cache mechanism

For operations in which a certain main storage boundary or a specific location must be observed, the Address Check Boundary (ACB) register and the /370 Address Compare register (AC) hold the appropriate address. These registers

Memory Management Unit Chip

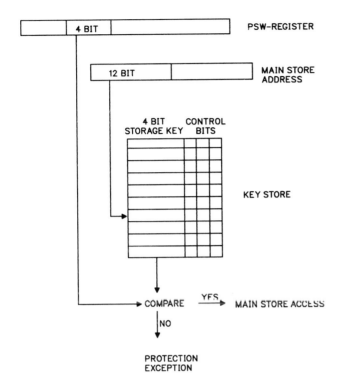

Figure 27. IBM S/370 key store

are loaded via the address register by specific bus commands (caused by CPU control microinstructions).

The S/370 address compare function causes the CPU chip to stop when the address preset in the AC register matches the address used in specified types of S/370 main store references. The MMU chip works with real and virtual addresses. For both addressing modes, the contents of the bus address register are compared with the contents of the AC register to detect any compare stop on either a real or virtual address.

The ACB register holds the 4 KByte block address of the boundary between non-S/370 memory and S/370 main storage. Both memories are addressed via 24-bit real addresses. The non-S/370 memory holds microroutines, I370 code, internal buffer space and control data. Checking logic ensures that memory operations directed to one address space do not access inadvertently the other address space. The ACB register contents are therefore compared with the contents of the address register during each real operation, and are compared with the TLB output during each virtual operation. The TLB supplies the real address in the latter case.

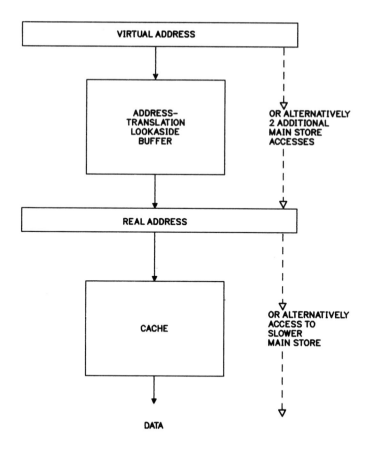

Figure 28. Serial access to high speed buffers

Address Translation occurs with the help of a 2-way set associative TLB, that stores 2 x 32 entries. Its structure is shown in Figure 24. Each entry holds a virtual/real address pair. A particular field of the virtual address is used as a pointer which selects a TLB entry (or row). Another field of the virtual address is fed from the address bus into the TLB compare logic where it is compared with the virtual address read out from the selected TLB entry. The comparison shows whether or not a translation exists for this virtual address.

The hit status is fed into the cache directory comparator which stops further actions if no hit has been achieved.

The cache directory indicates whether or not the addressed data is available in the cache. To do this, the directory holds a 15-bit entry (plus two control bits) for each of the 64-byte cache lines that reside in the cache. Such a cache directory entry (address portion) is read out and compared against the real address which is supplied from the TLB entry. The result of this comparison is a 1-of-4 code which points to one of the four cache compartments. This code is used when the TLB has signaled a "hit."

Memory Management Unit Chip

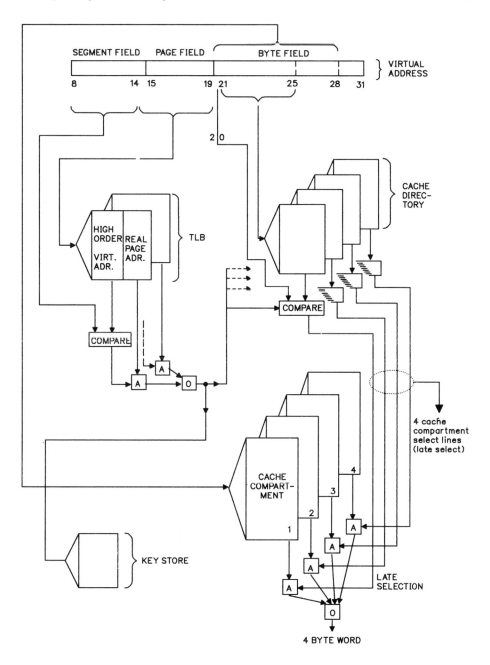

Figure 29. Simultaneous access to TLB, cache, and cache directory

For a store operation into the cache storage, the contents of the data register are aligned (on any byte boundary) prior to the actual store operation. Alignment ensures efficient operation and is necessary because the S/370 architecture

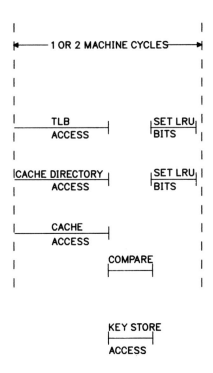

Figure 30. Parallel buffer access

is byte-oriented, while the Capitol chip set operates with fullwords to increase efficiency.

Fetch Operations by the CPU chip require to align the data in the fetch aligner prior to placement on the Processor Bus. This alignment ensures that the CPU chip gets the addressed data with the low order bit position truly located in the low order place (as necessary for arithmetic operations).

Most access operations to main storage also result in the automatic addressing of the key storage which is part of the MMU chip. The storage protection key itself is part of the current Program Status Word (PSW) stored on the CPU chip. The key read out of the key store is compared with the key field in the PSW, which is transferred over the key status bus, a part of the Processor Bus. The compare result permits the main store access either to proceed in normal fashion or results in an access exception indication.

Memory Management Unit Chip

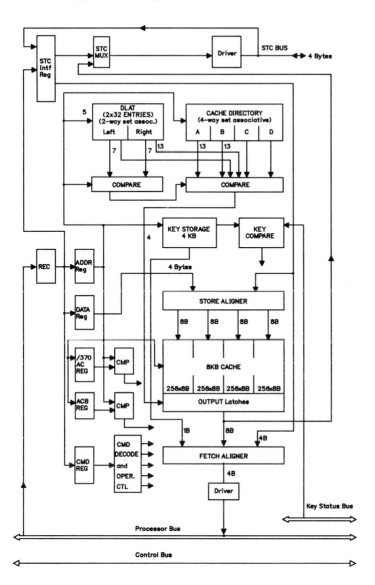

Figure 31. MMU chip data flow

2.4.4 Array Macros

Identical multiple array macro designs implement both the TLB and Cache Directory. Their realization is described in "4.5.1 Array Configurations" on page 218. Each is organized as 4 x 32 x 17, i.e. it holds 4 x 32 entries each 17 bits wide. The four subarrays are controlled by 4 independent Read-Write lines and each has separate Data In and Data out pins. One common Select clock sets the array in its active state. This organization allows a random access to any subarray. Both the TLB and cache directory are "set associative" implementations, as described in "2.4.2.2 Translation-Lookaside Buffer" on page 51. The TLB is implemented "2-way set associative", the Cache Directory "four-way set associative". Both sides of the TLB and all four compartments of the cache directory are accessed simultaneously.

Cache and Key storage have different organizations and different designs. All 4 arrays have uni-directional input and output lines.

2.4.4.1 Translation-Lookaside Buffer (TLB)

The TLB is a virtual address translation facility which uses this array macro in a 2 x 32 x 34 organization. Each entry requires 34 bits. Since the array macro has only 17 bits, 2 macros are always accessed in parallel to obtain 34 bits. This is achieved by grouping the 4 Read-Write lines in pairs of two.

The TLB has the structure shown in Figure 24. The subdivision into individual field is shown in Figure 32. The row- or entry selection is done by decoding bits 15..19 of the actual virtual memory address.

Virtual addresses are 24 bit wide. Address bit number 8 is the most significant bit, number 31 the least significant. Bits 0..7 are unused (when transmitting an address over the Processor Bus, they are used to transfer a command byte). Please note the S/370 architecture numbers bits different from most other architectures; bit position 0 contains the most significant bit, bit position 31 the least significant bit. A 24 bit address occupies bit positions 8...31; bit position 0...7 are unused.

The S/370 virtual address comprises bits 8..19 of the 24 bit address range. Bit 15..19 are used to address one of the 32 TLB rows with its two parallel entries. Each entry holds bits 8..14 of a virtual address and bits 8..19 of the corresponding real address. The two virtual address parts are read out and compared against the virtual address part of the virtual memory address.

Each TLB side contains 7 bits related to the storage key: 4 architectured /370 storage key bits - referred in Figure 32 as K, the Fetch protection bit, the Reference bit and the Change bit. When updating the TLB these bits are copied from the key store into the TLB. For each mainstore access, the 4 storage key bits are compared against the associated storage key transmitted over the Key-Status-Bus.

The Dynamic Address Translation (T) bit is part of the Program Status Word that is maintained on the CPU chip. It is transmitted (see Figure 39) as Key_status_bus bit 4 of the memory command, and compared against the T-bit stored in the TLB. It can be considered as an additional address bit.

A TLB hit is valid if there is a threefold match of the virtual address, the key access bits and the T-bit and if the valid bit is one. In case of a hit on one side these real address bits are fed (among others) to the cache directory comparator. The Validity (V) bit in Figure 32 indicates whether an entry in the appropriate TLB side is valid. It is set to one by the command "update TLB entry" and set to zero with the command "Invalidate TLB Row". The

Memory Management Unit Chip

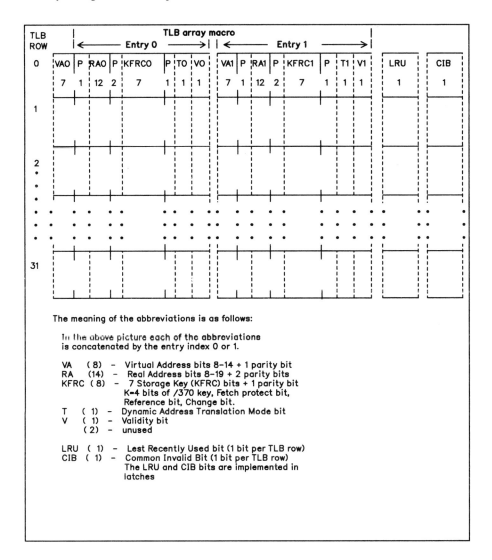

Figure 32. TLB structure

commands are generated by corresponding CPU control microinstructions ("2.2.6.2 Microinstruction Types and Formats" on page 34) and transferred to the MMU chip via the processor bus in bit positions 0..7.

The 32 Common Invalid Bits (CIB) are not part of the array macro. For performance reasons the control microinstruction "Invalidate TLB CIB" must execute in one cycle. As the array macro is not able to reset a certain bit in all 32 or 64 entries simultaneously, this function (equivalent to a "Purge TLB") had to be implemented in latches.

For a similar reason the LRU bits are implemented in latches. The LRU bit decides which side of the TLB is to be replaced in the case of a TLB miss. There are 32 bits associated to the 32 pairs of entries in the TLB. LRU = 1 means: last access was on side 0, the next update will be on side 1, and vice versa. This bit is individually flipped for each entry with each virtual access or with a TLB update instruction. The LRU bit cannot be part of the array macro because it must be written in the same cycle in which the corresponding TLB entry is read.

Both sides of the array macro feed the virtual address bits 8..14 in parallel into the TLB **compare logic**, where these bits are compared to the virtual address of the actual memory operation. A match is indicated by a TLB-Hit-0 or TLB-Hit-1. A checking mechanism stops further execution if erroneously both Hits get active. The compare logic is speed optimized by using fast CMOS circuits and by maintaining short wiring distances.

In parallel to the virtual address compare the array feeds the real address bits 8..19 of both sides into the compare logic of the Cache Directory. (See Figure 29). There these bits are padded with address bit 20 from the actual memory address and are compared with the real address bits 8..20 of all 4 compartments.

Several **TLB commands** manipulate data stored in the TLB. The "Invalidate TLB CIB" commands sets all 32 Common Invalid Bits to 1. The CIB bits are all stored in latches (as opposed to the TLB array) to facilitate quick command execution. The "Invalidate TLB Row" command selectively resets TLB entries. The command bits 8-19 define a virtual 4 K Byte page address, bits 20-31 are not used. Bits 15-19 of this virtual page address are used by the MMU to select a TLB row using the set-associative addressing scheme. Both TLB entries in the selected TLB row are invalidated by setting the validity bits V0 and V1 to 0.

The "Update TLB Entry" command is used to change TLB entries. The execution of this command requires 3 machine cycles. In the first cycle bits 15-19 of the virtual 4 KByte page address are used by the MMU to select a TLB row. The TLB LRU control determines the TLB entry to be updated. The dynamic address translation bit of KEY_STATUS_BUS_4 is compared with the T-bit in the appropriate TLB entry. In the second cycle the CPU chip puts the corresponding real 4 KByte pag e address on bit positions 8-19 of the Processor Bus. The remaining bits are not used. In the third cycle the real 4 KByte page address is written into the real address (RA) part of the TLB entry. Bits 8-14 of the virtual 4 KByte page address are also written into the virtual address (VA) part of the selected TLB entry. The Dynamic Address Translation bit is set into the T bit of the selected TLB entry. Finally, the MMU sets the valid (V) bit to 1 in the selected TLB entry. The valid (V) bit in the TLB entry not selected is set according to CIB. If CIB is 1 then V and CIB are set to 0; otherwise, V remains unchanged. Using the real 4 KByte page address the MMU sets the reference bit in the key storage to 1 and writes the updated storage key into the KFRC part of the selected TLB entry.

2.4.4.2 Cache Directory

The Cache Directory uses the same array macro realization as the TLB, but in a 4 x 32 x 17 organisation. It is 4-way set associative thus forming 4 "directory compartments" with 4 associated compartments in the cache array itself. Each Cache Directory compartment holds 32 real address entries (bits 8..20 of the real address). Each entry assigns a line in the cache array.

The Cache Directory has the logical and physical structure shown in Figure 26. The field assignment in each cache directory is shown in Figure 33.

The row- or entry selection is done by decoding real address bits 21..25. The entry contains in each compartment real address bits 8..20. All four compartments are compared in parallel with the corresponding bits fed from the TLB (bits 8..19) which are complemented by address bit 20. A compare hit indicates that the appropriate cache line (address bits 8..25) of 64 bytes is available in the cache array. This line is located in the cache entry determined by start address bits 21..25 and bits 26..28 equals zero.

Two control bits are associated to each Cache Directory compartment. The Validity bit indicates whether an entry in the appropriate Cache Directory compartment is valid. It is set to zero by the command (control microinstruction) "Invalidate Cache Directory Entry" and set to one after the completion of a cache line load procedure.

The Change bit indicates whether a valid line in the cache was changed by a store operation and whether as a consequence a cache line must be stored back to main store before a new cache line is transferred to the cache. It is set to zero when a new line is transferred to the Cache.

Figure 34 shows the Cache Directory compare logic (see also the right side of Figure 29). There are three sources of the real address: TLB side 0 and TLB side 1 in the virtual addressing mode. (In this case the real address bit 20 from the bus input register is concatenated to the TLB real bits 8..19). The third source is the bus input register in the real addressing mode, where the TLB is bypassed and where the bus input register bits 8..20 are compared to the corresponding cache directory bits. For performance reasons all three sources are compared simultaneously to all 4 compartments of the Cache Directory. Also this logic is speed optimized in the same way as at the TLB compare logic.

The compare results of all 12 compare logic parts are gated according to the three address sources. This applies in case of a cache hit. A forth gate condition is included for a cache miss. Here, in order to perform a line load or cast-out operation the contents of the LRU registers generate the cache compartment select lines.

Only one of the 4 cache selects may point to one of the four cache array compartments. A checking mechanism stops further execution if erroneously more than one select line is active. The cache compartment select lines are called "late selects" for the cache, because this is time wise the latest decision which of the four columns is enabled to drive its data in the Fetch Aligner. To resolve a

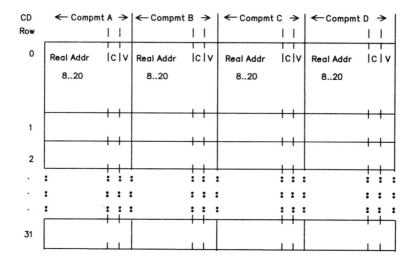

Figure 33. Cache directory structure

cache miss situation, the MMU sends a line-load operation to the STC chip to fetch 64 Byte from main store in burst mode. The cache directory maintains a LRU scheme to determine which entry in a cache row was "Least Recently Used" and thus is subject to replacement. "Used" has the meaning of "referred to". Or, if the line was changed since loading, a cast-out command must be executed to store data from cache back to main store before the new line may be fetched.

A six bit register for each of the 32 Cache Directory entries points to those compartment in that entry which was not referred to for the longest period of time. Each reference to an entry changes the code word in a certain way.

A six bit register was chosen for the following reason. Let 1, 2, 3, 4 be indicators for the sequence of reference to a given compartment; 4 = least recently used, 1 = last recently used. At a given state of the system each of the 4 compartments may have any indicator value. The total number of permutations is 24. To code 24 events 5 bits are required. To make the scheme more transparent and to allow a simple coding and encoding algorithm a sixth bit was added (see Figure 35).

Memory Management Unit Chip

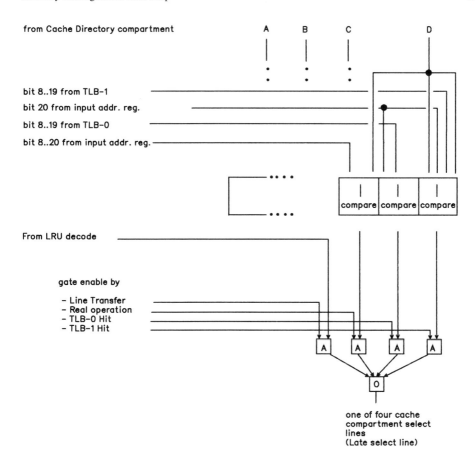

Figure 34. Cache directory compare logic

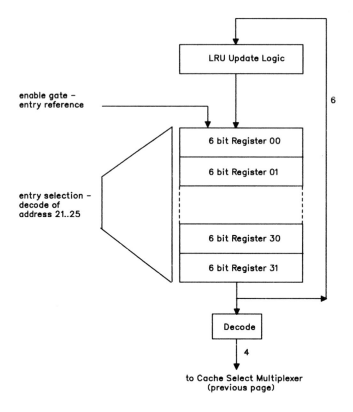

Figure 35. LRU logic schematic

2.4.4.3 Cache Array

The Cache Array provides a fast access to 8 KByte of data or instructions. It is organized as 4x256x8 Bytes, i.e. 4 compartments of 2 KByte are accessed in parallel (see Figure 36). Each compartment has 256 entries, each entry is 8 Byte wide. Eight adjacent entries form a 64 Byte cache line. The entry selection is done by decoding real address bits 21..28. (The cache has a One-byte-write capability, provided by an independent read/write line for each byte within a compartment entry. These 8 R/W-lines are common for all 4 compartments.)

For a fetch operation three arrays are addressed simultaneously: virtual address bits 21-25 address the four columns of the cache directory, bits 21-28 address the cache data array and bits 15-19 address the TLB (see Figure 29). Key-controlled protection checking is performed using the storage key in the selected TLB entry. The same addressing scheme applies, for a store operation except that the proper cache column must be selected before the data can be stored. Data is aligned (on any byte boundary) prior to the actual (partial) store operation.

Memory Management Unit Chip

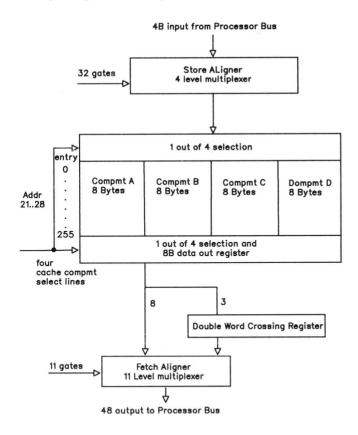

Figure 36. Cache structure

In a S/370 instruction- or data stream an instruction or a data string frequently crosses a logical 4 Byte boundary. This fact should not impact the Processor performance by requiring a second cache access. Therefore the data width of the cache is 8 Bytes. It has an 8 Byte internal data out register. If the less frequent double word crossing (crossing of an 8 Byte boundary) occurs a second cache access is done and the data of the first access is stored in the **Double Word Crossing (DWC) Register** It is three bytes wide and contains either bytes 0..2 or 5..7 of the first cache access, depending on a decrement or increment operation. The correct alignment of the content of the Cache output register and the Double Word Crossing register is done in the Fetch Aligner. It properly aligns the data on the Processor Bus. This is necessary, because the S/370 architecture is conceptually byte oriented. The Capitol chip set physically operates on a fullword basis for performance reasons. An incrementing/decrementing logic in the bus address register generates the correct cache address bit 21..28 for each data read or write operation which is controlled by one or more of the 8 R/W lines.

A 64 Byte line is brought from main store via the Store Aligner into the cache by 16 consecutive data transfers, as the interface between cache and main store has a 4 Byte width. In this case the R/W lines are activated alternatively in groups of 4 (0-3 or 4-7) depending on odd or even data transfer from main store.

A read access of the cache provides 8 bytes of data in increasing address sequence. The CPU chip is byte oriented and able to access data in an increasing or decreasing addressing mode. Its internal bus is 4 Byte wide regardless of boundary limitations with the physical storage hierarchy.

Therefore it is necessary to place data on the Processor Bus with the alignment required by the CPU chip. This alignment could be done by a slow shift register operation after the data are available at the cache output. Instead, for performance reasons, the Fetch Aligner is implemented as an eleven level multiplexer. The appropriate multiplexer gates are activated in parallel to the cache directory- and cache access based on start address and length count of the memory fetch command and based on the additional interface line from the CPU which determines Increment/Decrement Addressing Mode. The eleven levels have to following meaning:

5 levels   adjust the 4 bytes of data within 8 bytes of the cache width if no double word crossing occurs;

3 levels   adjust the 4 bytes of data within 8 bytes of the cache width and 3 bytes of the Double Word Crossing (DWC) register if double word crossing occurs in incrementing address mode;

3 levels   adjust the 4 bytes of data within 8 bytes of the cache width and 3 bytes of the DWC register if double word crossing occurs in decrementing address mode;

Actually two more levels are implemented: One for data transfer from the main memory to the Processor Bus and one for reading the Keystore and some chip internal status registers.

If the length count is less than 4 bytes, the left most bytes are forced to zero on the Processor Bus.

The Store Aligner has a similar function as the Fetch Aligner. It assures correct data byte placement from the Processor Bus into the Cache.

It consists of 8 multiplexers, each with 4 levels, and aligns each of the 4 bytes from the Processor Bus to each of the 8 positions in the cache. In addition the Store Aligner has two more levels, one for testing the cache array and one for storing aligned data from main memory into the cache in case of a cache miss (line load operation).

As with the Fetch Aligner, activation of the 32 multiplexer gates occurs in parallel to the cache directory- and cache access, based on start address and length count of the memory store command and based on the CPU interface line "ADDR DECREMENT" (see Figure 86).

2.4.4.4 Keystore

The S/370 storage protection facility consists of a key store with a 7-bit storage key entry for each 4 KByte block of S/370 main storage. (Older versions of the S/370 architecture used to support a 2K Byte storage protection block as well; the Capitol chip set does not implement this)

The key store is physically implemented as a 4K x 8 array on the MMU chip. Each entry has 7 functional bits plus one parity bit. Figure 37 shows how the key store is embedded in the multiplexer- and compare logic circuitry, the connections to the TLB, which contains copies of the key array, and to the outside buses. The main purpose of the Auxiliary_Key_Register is to modify the Reference- and Change bits in the Key Array and in the TLBs. The entry selection is done by decoding real address bits 8..19. The KEY MISMATCH condition is signaled to the CPU chip via the Key_Status_Bus if a S/370 protection violation has occurred, detected either by a key store mismatch or by Key information stored in one of the TLB sides. The 7-bit storage key layout is shown in Figure 38.

To allow S/370 key-controlled checking at the beginning of a main store access, the 4 storage key bits are transferred in parallel to the memory address over the dedicated, parity checked Key_Status_Bus. The Dynamic Address Translation (T) bit in the PSW is also transferred over this bus. It is a bi-directional bus with 5 data lines and 1 parity line. In addition to this, the Key_Status_Bus has the dual function of transmitting operational status bits from the BBA, FPU, and MMU chips to the CPU chip. The bus layout is shown in Figure 39.

Storage access requests normally originate from the CPU or BBA chip. If the CPU chip requests access, the storage key normally originates from the current PSW. If the BBA requests access, the storage key normally originates from the S/370 Channel Address Word (CAW) used for the current I/O data transfer.

The Dynamic Address Translation (T) bit defines whether a real or virtual address is used in the command. For CPU main store accesses it is normally taken from another field in the S/370 current PSW. For I/O-memory commands it must be 0, indicating that a real address is on the bus. It is compared against the T-bit in the appropriate TLB entry. If it does not match a TLB_MISS condition is signalled on the Status_Bus.

Special microcode control instructions are used to generate Processor Bus commands that modify key store data. They read and write key store entries, and modify the reference and/or change bits.

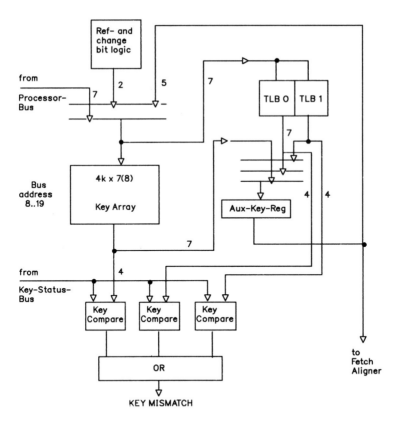

Figure 37. Keystore structure

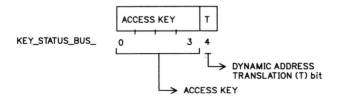

Figure 39. Key_status_bus layout

2.4.5 Storage Controller Interface

The MMU chip interfaces to the STC chip which handles the interface lines to the memory cards that implement main store. It is described in "2.5.2 Memory Organization and Control" on page 81.

A 4 Byte bi-directional bus plus some control lines interconnect the MMU and STC chips. A five level multiplexer provides the main part of the sending logic and a 4 Byte register decouples the timing of both chips. The connection of this

Memory Management Unit Chip

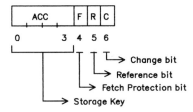

Figure 38. S/370 storage key

interface logic to other parts of the MMU functions is shown in Figure 40. Two types of operations are executed over this interface.

One type are the cache line load - and cache cast out operations. If a cache miss condition is detected the cache control logic generates an appropriate command and calculates the (real) main store address where the desired data are located. Command and address are fed via the STC_Multiplexer over the STC Bus to the Storage controller. It responds by sending 64 bytes of data in 16 consecutive transfers of 4 bytes each. The STC_Interface_Register decouples the STC Bus from the internal cache input bus, which ends in the Store Aligner.

Cast out operations are handled in a similar way. Command and address are transferred via the STC-Multiplexer to the Storage Controller. The data from the Cache output register use two multiplexer levels of 4 Bytes each. The 8 Byte cache output register decouples cache internal timing conditions from the STC Bus.

The second type of operations are the real Storage operations under the control of the CPU chip or of the BBA chip. In both cases the MMU chip is not logically involved and merely passes through data and command. The CPU can fetch blocks of 64 Bytes of microinstructions from non-S/370 memory. The BBA chip may fetch or store between 1-64 Byte of I/O data. The command and data in case of store operations from the Processor Bus are buffered in the STC-Interface-Register and fed via the STC-Multiplexer to the STC Bus. In fetch operations the Storage controller sends data which are again buffered in the STC-Interface-Register and than fed to the Processor Bus via the Fetch Aligner.

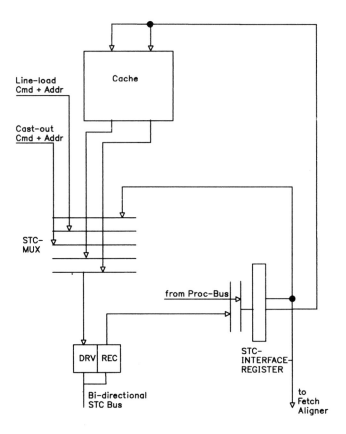

Figure 40. Interface to storage controller

2.5 Storage Controller Chip

Rolf Müller

The Storage Controller (STC) constitutes the interface between the MMU chip and the dynamic main memory. It manages the accesses to the memory and maintains the accuracy of memory data. The STC is a single CMOS chip connected to the cache controller and the memory via bi-directional 4 Byte busses and two sets of control lines. It performs complete memory handling with address generation, timing control, error correction, refresh scrub, redundant bit handling and memory initialization. Memory sizes ranging from 1 to 16 Mbytes, up to four memory cards and multiple types of memory cards are supported.

The STC chip uses 17299 cells (approximately 14K gates), and is packaged on a 36 mm ceramic Pin Grid Array (PGA). The chip size of 8.5 x 8.4 mm/2 is determined by the number of I/O pins needed.

Figure 41 shows the interconnection of the STC chip within the Capitol chip set.

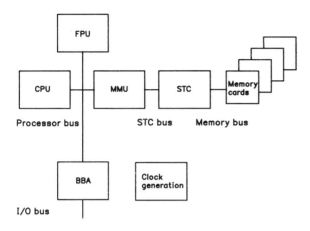

Figure 41. System structure

2.5.1 STC Chip Structure

The major units of a storage controller are the "Memory Control" unit and the "Error Correction" unit. Most storage controller implementations require separate chips for these units and the memory control unit is sometimes subdivided into "Memory Control" and "Timing Control", whereas the STC is a single chip implementation.

2.5.1.1 Memory Control Unit

The Memory Control unit (Figure 42) interfaces with the MMU chip via the 4 Byte STC Bus. Commands on this bus are decoded and appropriate memory control signals are generated to perform the commands. In addition to these external commands refresh, scrub and memory initialization is performed without CPU chip support. The main tasks of the memory control unit are command decoding, address translation, memory control signal generation (RAS, CAS, BS ...), timing control for memory and error correction unit, refresh/scrub interval counting, refresh/scrub address generation, and memory initialization.

In the case of a valid STC command on the STC Bus the **select decoder** activates the internal select signal. The valid-signal indicates that the information on the STC Bus is any command, and the cancel signal is used to cancel a selection. Cancelling of a command occurs if, in case of a cache hit, the command will be performed by the MMU chip.

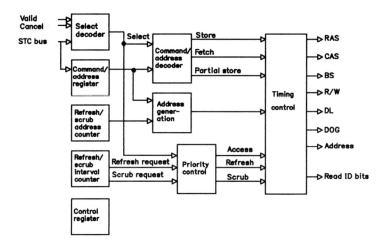

Figure 42. Memory control unit

The STC Bus is 4 Byte wide, with the first byte defining the command and the last three bytes defining the address (see Figure 43). Valid commands interpreted by the **command / address decoder** are STORE, FETCH, FETCH SLOW, EXTERNAL SCRUB and MESSAGE commands. STORE/FETCH from 1 to 64 Byte starting on any byte address is allowed. FETCH SLOW is a 64 Byte fetch command with a reduced data rate for loading of the control store. Message commands deal with setting/sensing/resetting of the STC internal control registers, check register and redundant bit directory.

Depending on the number of bytes to be stored and the address, a store command may result in a partial store operation (a store of less than four bytes) requiring prefetch.

Figure 43. Command / address format

The MMU chip issues the start address of a fetch/store command and the length of the data transfer (number of bytes to be fetched or stored). The 24 bit start address, allowing addressing of 16 MByte, is converted by the **address generation** logic into memory control signals (RAS, CAS etc.) selecting banks of 1 MByte each, and a 9 bit row/column address for addressing within these banks. Furthermore the memory control signals and the row/column address have to be updated during data transmission.

The **refresh/scrub interval counter** generates refresh requests and scrub requests according to the required refresh interval stored in the control register. The 22 bit **refresh/scrub address counter** generates the memory addresses to be refreshed or scrubbed. It is also used as an address counter for memory initialization.

The **priority control** logic arbitrates the STC-generated refresh/scrub requests and the regular memory accesses generated via the STC Bus. The **timing control** generates the signal timing required by the memory cards. The **control register** is set during personalization and holds information like memory card type, memory configuration, memory size, refresh mode, refresh interval, enable scrub, and read memory card identification bits (ID bits).

2.5.1.2 Error Correction Unit

The Error Correction Unit (Figure 44) manages the data flow between the main memory and the MMU chip, performing parity check, error detection/correction, merging of data in partial store mode etc.. It consists of three registers, 39 or 40 bits wide, the error correction/parity check part, a control register, a check register and the redundant bit directory.

The data flow for basic operations is explained below. For a description of error correction see section 2.5.3.

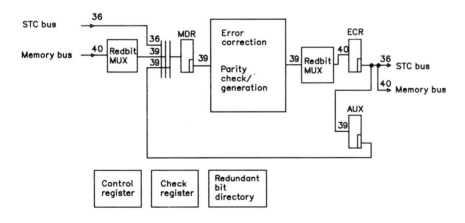

Figure 44. Error correction unit

For a store operation the data bits from the STC Bus, to be stored in the memory, are latched in the MDR register. After parity check and generation of the check bits the defect bit of the addressed memory bank is multiplexed on the redundant bit (see "2.5.3.4 Redundant Bit" on page 93), and data and check bits are latched in the ECR register. Afterwards the 32 data bits, 7 check bits and the redundant bit are placed on the memory bus.

For a fetch operation the 40 bits from the memory bus are reduced by the redundant bit multiplexor (REDBIT MUX) to 32 data bits and 7 check bits and latched in the MDR register. Afterwards error detection/correction is performed and in the case of a double error a retry request to the memory control unit is issued. The corrected data and the parity bits are latched in the ECR register and finally placed on the STC Bus. If the retry procedure was not successful or disabled, a check line will be activated and error type as well as the address of the non-correctable error are latched.

A partial store operation is a specific store requiring prefetch and merging of data. The memory bus is four bytes wide so that check bits are generated for every four bytes of data. If, however, less than four bytes are to be stored, a prefetch of four bytes with error correction is performed, and the corrected data is latched in the AUX register. Then the bytes to be stored from the STC Bus are merged with the bytes of the AUX register and latched in the MDR register. The remaining part is identical with normal store operation.

The **control register** is set during personalization, and holds information used for diagnostic purposes, e.g. inhibit ECC, disable retry, test ECC, force error, check bit read out, and force parity error. For details see section 2.5.4.

The **check register** is set if an error condition (parity error on the STC Bus or non-correctable error during memory fetch operation) occurs. It holds information like uncorrectable error (UCE), soft/soft error, hard/hard error, UCE address, parity error on the STC bus, Byte position of the parity error, and parity error in command/address or data.

The **redundant bit directory** is a 16x6 bit register that holds the positions of the defect bit per bank (see "2.5.3.4 Redundant Bit" on page 93) and controls the two redundant bit multiplexors.

2.5.1.3 STC Chip Interfaces

The STC Bus (interface between the MMU and STC chips, see Figure 41) is a bidirectional 36 bit bus (32 command/address or data bits, 4 parity bits) used for command/address and data transmission. There are 7 additional control or check lines.

The main memory interface consists of the bidirectional 4 Byte memory bus and 43 control lines including 9 address lines. The clock interface consists of 7 clock lines (two master clocks, three slave clocks and two shift clocks, see "2.9.1 Clock Signal Types" on page 118).

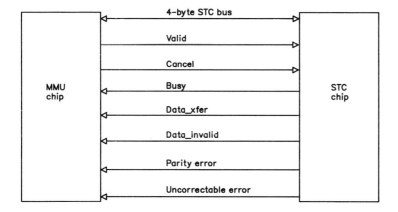

Figure 45. Cache controller interface

2.5.2 Memory Organization and Control

2.5.2.1 Overview

The STC chip is designed to communicate with IBM's standard memory cards that are used in many products, e.g. the IBM AS/400 system. Two basic types of memory cards are supported, using 265 Kbit chips or 1 Mbit chips. These memory cards are organized in "banks" with a size of 1 MByte and range from 1 MByte to 16 MByte. The STC chip can support memory sizes of 1,2,4,8,12 and 16 MByte distributed on up to four memory cards. For example a memory of 4 MBytes may be configured with four memory cards of 1 MByte each, two memory cards of 2 MBytes each or a single memory card of 4 MBytes.

Handling of these multiple configurations requires a specific board wiring for the four memory cards and an identification procedure after power-on to adopt the STC to the memory. The memory cards offer identification bits accessible via the data bus. In this way information on memory card type, memory card size, and in general insertion of a memory card can be obtained.

2.5.2.2 Memory Card

The following describes an 8 MByte card as one of several implementations. It is organized in four "blocks" of 2 Mbytes each, see Figure 47.

Each block is assembled of twenty 1-Mbit chips. The memory chips are x4-organized, thus a minimum of ten chips is needed to generate a 40-bit memory bus (32 data bits, 7 check bits, 1 redundant bit, see "2.5.3 Data Integrity" on page 85). However, an organization with only ten chips feeding one word (32 data bits plus ECC bits) creates a significant problem. In the case of a "chip kill" (a chip is totally bad on all four outputs, which is an error condition with a relatively high probability), a 4-bit error appears in a single

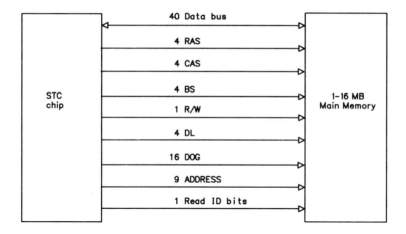

Figure 46. Main memory interface

word. Since with seven check bits only single and double errors (SEC/DED) can be detected, this chip kill might be undetected. For this reason, the memory card is organized such that the four outputs of a chip belong to two different words, making chip kills detectable.

In order to economize on cost, the array chips on the memory cards minimize the number of pins on each module. Towards this end, the address lines into each chip are time multiplexed. First one half of the address lines are stored in latches on the chip with the help of a Row Access Strobe (RAS) signal. Subsequently, the chip receives the second half of the address lines with the help of a Column Access Strobe (CAS) signal. The term RAS as used has a different meaning than the acronym RAS standing for Reliability, Availability, Serviceability (RAS) as used in "2.11 Reliability, Availability, Serviceability" on page 132.

Each of the four memory blocks has its own RAS and CAS controls but shares R/W (read/write) and BS (bank select) controls, plus address and data registers for the 2x40 DI/DOs (Data In / Data Out) with another one. These data registers are used in fetch and store operations and set by four DL (Data Latch) signals. Four DOG (Data Out Gate) signals control the connection of the data registers to two data busses of forty bit each and enable the bus drivers.

Storage Controller Chip

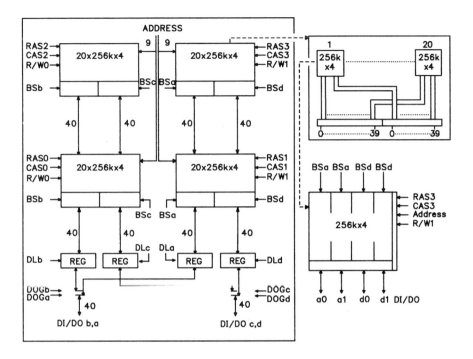

Figure 47. 8 Mbyte memory card organization

2.5.2.3 Memory Performance

Two key parameters define the performance characteristics of a memory: first access, and data rate. First access is the time from activating RAS to valid data on the memory bus (8 machine cycles in this implementation) whereas data rate indicates the speed at which consecutive words after the first one can be fetched from the memory. While first access is limited by the speed and drive capability of the memory chips and drivers and the capacitive loads resulting from wiring and fan-out, data rate can be optimized by special features.

To achieve a data rate higher than that achievable with consecutive first accesses, memory chips feature a number of specific modes, e.g. nibble mode, ripple mode, page mode, static column mode. Besides these modes offered by the memory chips, memory interleaving is another way to increase the data rate. The combination of page mode and interleaving is implemented and described.

Page Mode is one technique that achieves a higher data rate. By activating RAS and applying a 9 bit row address that corresponds to the word lines (Figure 51) as many cells as bit lines (in this specific case 512) are connected to sense latches. By subsequently activating CAS and applying a 9 bit column address that corresponds to the bit lines, just one sense latch, or four sense latches in a x4-organized memory chip, are connected to the DI/DOs. The

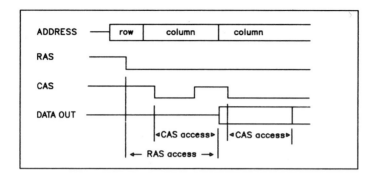

Figure 48. RAS / CAS access

time interval from activating RAS to the appearance of the data at the DI/DOs is the first access or RAS access (see Figure 48)

If only a single access to the chip has to be performed, CAS and RAS will now be deactivated and the read operation is terminated. If, however, consecutive read operations on the same row address have to be performed, application of consecutive column addresses, indicated by toggling CAS, will fetch all the information from the sense latches that were also set by RAS. Because the information to be fetched is already stored in the sense latches, these fetches have a shorter access time (CAS access).

Interleaving is another technique to obtain a higher data rate. If a memory is organized in a way that consecutive words are located on different chips, it is possible to access these chips in a staggered mode (interleaving), thus increasing the data rate. The interleaved groups of chips are called banks and have a size of 1 Mbyte. Depending on the memory size the STC performs no interleaving or interleaving between two or four banks (2-way/4-way interleaving). It interleaves between banks on the same memory card or on up to four different memory cards. The maximum continuously data rate of four bytes per cycle is achieved with 4-way interleaving. Two-way interleaving results in a one-cycle gap after every four bytes of data whereas transmission without interleaving will result in three gap-cycles after every four bytes of data (see Figure 49).

2.5.2.4 Fetch Operation

Figure 50 shows the principle timing of a 64 byte Fetch operation for an 8 MByte memory card. Note that all signals are low active.

RAS0 and RAS1 select two blocks of 2 MByte each (four banks), and define the row address according to the address bus. In both blocks the same row address is selected. Then CAS0 is started selecting columns in the left block according to the column address. The two words selected by CAS0 are set on the memory bus with DOGb and DOGc. Further words are fetched from the left block with consecutive cycles of CAS0 in page mode. The gap in the

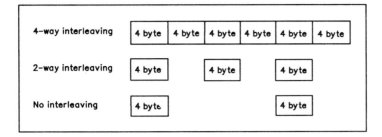

Figure 49. Data transmission vs interleave modes

stream of data after every two words (8 bytes) from the left block is filled by words from the right block selected by CAS1, DOGa and DOGd. Thus, by interleaving between two 2 MByte blocks (four banks), the data rate on the memory bus is twice that of a single block. DI/DO shows the sixteen words of four bytes each. The described 64 Byte FETCH takes 25 cycles.

2.5.2.5 Store Operation

Figure 52 shows the principle timing of a 64 Byte Store operation.

The data to be stored is set on the memory bus with a data rate twice that of the memory chip's page mode, and set in the registers with staggered DLb,c,d,a. As described in 64 Byte FETCH the two 2 MByte blocks are running in page mode and are interleaved by the staggered CAS0, CAS1. Contrary to the fetch timing, BS is also timed to make sure that only two out of four data inputs of the memory chips are set at a specific time. The described 64 Byte STORE takes 21 cycles.

2.5.3 Data Integrity

There is a small but existing probability that some of the dynamic memory cells may lose their information. Since this risk increases with the size of the memory, larger memories require an additional effort for data integrity. In addition to error detection and correction, provisions have to be made to prevent errors from accumulating.

Two types of errors have to be considered: hard errors and soft errors. Hard errors result from physical damage. When read, a cell will always provide the same information, regardless of what was written to it. The cell is "stuck at one" or "stuck at zero". Soft errors are a temporary malfunction. A cell loses its information, e.g. resulting from alpha particles, noise or pattern sensitivity. After rewriting, the cell will work properly again.

In addition to error detection/correction, refresh of the dynamic memory and indication of address faults are part of data integrity. Other features dealing with data integrity are Complement Retry, Redundant Bit, and Scrub.

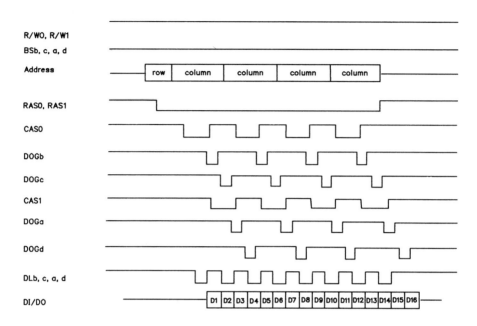

Figure 50. Fetch 64 byte

2.5.3.1 Refresh

Two types of semiconductor memories using "static" or "dynamic" storage cells, are used in computer main memories. They differ in the way the information, "1" or "0", is stored. For static memories this cell is a flip-flop that, after setting it to any of the two steady states, will stay in this state as long as it is not reset to the opposite state. In a dynamic cell the information is stored in the charge of a small capacitor. A logical 1 is represented by charging the capacitor to the supply voltage, whereas a logical 0 is represented by a discharged capacitor. However, due to leakage current a capacitor will lose charge, so that after a certain time a logical 1 will become a logical 0.

To avoid this loss of data, dynamic cells have to be continuously refreshed. They have to be recharged to the maximum voltage level before the voltage drops to a level no longer representing a logical 1. The time between these refresh cycles (data retention time) is about 4 ms.

A complete read operation followed by a write operation is one possibility to perform the refresh operation. More efficiently, a refresh is performed by merely connecting the cell to be refreshed to a sense latch (Figure 51). This sense latch detects the level of the cell, even if it is just a "weak" logical 1 or 0, and switches to the appropriate state, thus recharging the cell to the maximum voltage.

A second feature that eases refresh is that not only one cell can be refreshed at a certain time, but as many cells as sense latches are available. By activating

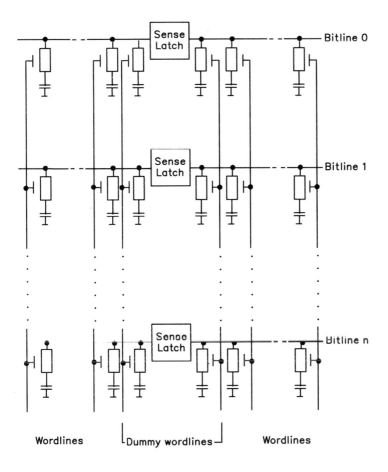

Figure 51. Memory chip organization

one word line, all cells connected to this line will be connected to "their" sense latch thus being refreshed simultaneously. The described behavior drastically reduces the overhead for refresh, so that the availability of the memory, i.e. the time where no refresh is to be performed, is in the range of about 97%.

The STC chip performs refresh independent from the CPU and MMU chips (hidden refresh), requiring six cycles for every refresh. An internal counter (refresh/scrub interval counter) issues refresh requests with a programmable interval of up to sixteen values. Every refresh request triggers a memory refresh immediately after completion of an eventually running memory access.

As described, refresh of dynamic memories requires merely activation of word lines, hence only RAS and a row address have to be applied. CAS and a column address are not required. The refresh row address is updated by an internal 9-bit refresh address counter. Depending on power dissipation limitations, either all memory chips can be refreshed simultaneously (full array refresh) or only half of them (half array refresh). In the first case all four RAS lines are activated simultaneously while in the second case only two of the four RAS lines are activated at a certain time. Selection of refresh interval and full/half array refresh is achieved via five bits in the control register. Full array refresh and a data retention time of 4 ms require a refresh interval of 7.68 μs.

If a regular memory access via the STC Bus is requested during a running memory refresh, the execution of the command will be delayed until completion of the refresh.

2.5.3.2 ECC (Error Correction Codes)

The length of a code (number of bits) depends on the required number of words to be coded. Code length and number of words follow the rule:

$$x = 2^n \quad (\text{n = number of bits, x = number of code words})$$

The idea of error detection is to use more bits than necessary, thus implementing redundancy in the code. The values of these added bits are defined via special algorithms making errors detectable. A key parameter for error detection/correction capability is the "distance" of a code. The distance is equivalent to the minimum number of bits by which valid code words differ from each other. A code with the minimum number of bits has a distance of one, because all possible combinations are used as valid code words. By adding bits to the code, implementing redundancy, the distance is incremented. A distance of one does not allow any error detection, a distance of two enables detection of single errors, a distance of three enables correction of single errors etc. (see Figure 53).

Storage Controller Chip

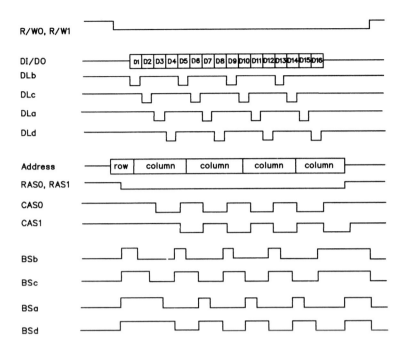

Figure 52. Store 64 bytes

Distance	Errors to be detected	corrected
1	0	0
2	1	0
3	1	1
4	2	1

Figure 53. Distance of codes

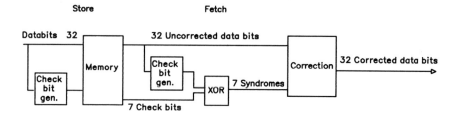

Figure 54. Error correction principle

Data bits	Check bits
11	5
26	6
57	7
120	8
247	9
502	10

Figure 55. Check bits vs data bits

	DATA			CHECK
Byte 0	Byte 1	Byte 2	Byte 3	bits
01234567	111111 89012345	11112222 67890123	22222233 45678901	0123456
11111111	11100000	00000011	00000001	X
00000000	11111111	00101101	00000101	X
11100000	10011100	01010110	00101010	X
10011000	01010010	10011000	01010010	X
01010110	00101010	11100000	10011100	X
00101101	00000101	00000000	11111111	X
00000011	00000001	11111111	11100000	X

Figure 56. Check bit matrix

Storage Controller Chip

```
1. Stored data        : 00000000 11111111 00000000 11111111
2. Stored check bits  : 0010100
3. Fetched data       : 00000000 11111111 00000000 11111110
4. Fetched check bits : 0010100                           └─Error
5. New check bits     : 1110110
6. Syndrome bits      : 1100010  ──→ - Unequal 0
                                     - Error
                                     - Matches column below
                                       data bit31
                                     - SE in data bit 31
                                     - Invert data bit 31
```

Figure 57. Single error detection / correction

```
1. Stored data        : 00000000 11111111 00000000 11111111
2. Stored check bits  : 0010100
3. Fetched data       : 00000000 11111111 00000000 11111100
4. Fetched check bits : 0010100                          │└─Error
5. New check bits     : 1101100                          └─Error
6. Syndrome bits      : 1111000  ──→ - Unequal 0
                                     - Error
                                     - No match with any column
                                     - Double Error
                                     - No correction possible
                                     - Retry request
```

Figure 58. Double error detection

Hamming distance-four codes are most commonly used and capable of Single Error Correction / Double Error Detection (SEC/DED). During store operation a number of so-called "check bits" is appended to the data bits. After reading the memory, "syndrome bits" are generated by comparing the fetched check bits via exclusive-or (xor) with newly generated check bits. If there is a mismatch, the syndrome bits will be unequal to "0", thus indicating the error. In the case of a single error the location of the error can be decoded out of the syndrome bits and a correction will be performed (Figure 54).

Note that this algorithm corrects all single-bit errors and detects all double-bit errors. Multiple-bit errors (more than 2) may be detected but not corrected, or misinterpreted as single-bit errors. The number of check bits to achieve a distance of four depends on the width of the memory bus (Figure 55). Because these data bit numbers are not commonly used for the width of a memory bus the next matching number of check bits is to be used, hence a data bus of 32 bits requires 7 check bits. Note that because of this mismatch there is some potential in the check bits for additional information, e.g. to be used to detect whether double errors result from the same memory chip (helpful for x2-organized memory chips).

The STC performs error detection/correction with a modified Hamming code and uses 7 check bits for 32 data bits. It is able to correct single errors and to detect double errors in any combination on the 32 data bits and 7 check bits. Single errors will be corrected by the ECC while double errors will trigger a retry request to the memory control unit. Multiple bit errors (more than two) may be detected but not corrected, or misinterpreted as single-bit errors. The condition of a floating memory bus however (all "1" in this application) will be detected as a non-correctable error.

There are various algorithms to create the check bits, defined by so-called "check bit matrices". The check bit matrix implemented is shown in Figure 56. A check bit is generated by "XOR"ing the data bits marked with 1. For example, check bit 0 is generated by xoring data bits 0,1,2,3,4,5,6,7,8,9,10,22,23 and 31. The location of the single error is identified by comparing the syndrome bits with the columns below the data bits. Examples for single error correction and double error detection are shown below (Figure 57 / Figure 58).

2.5.3.3 Complement Retry

Implementing correction of double errors by the ECC would increase the number of check bits to thirteen. This unacceptable number led to the COMPLEMENT RETRY approach able to correct some of the double errors by a special fetch/store routine with no additional check bits.

As explained above, there are two types of errors: soft errors and hard errors. The example in Figure 59 explains the complement retry routine for a hard/hard error combination. The stored data (1) differ from the fetched data (2) by two inverted bits due to the two hard errors on bit 30 and 31 (bit 30 is stuck at 1, bit 31 is stuck at 0). The fetched data is complemented and restored (3) and afterwards fetched again (4). Now the fetched data is complemented again (5) and the resulting bit string is identical to the originally stored data. This example shows that with the complement retry feature any number of hard errors can be corrected. The disadvantage of the procedure is that, contrary to the ECC, it requires multiple fetch/store cycles but on the other hand it does not require additional check bits.

The example shows the ability of the complement retry feature to correct any number of hard errors. The problem however is to know whether there are errors in the bit string. Running this procedure with every memory fetch would destroy correct information if the hard errors match the stored data (check the procedure with 01 instead of 10 for data bit 30,31). Thus, the application of the complement retry feature is useful only in conjunction with ECC. The ECC checks for errors and corrects single errors immediately. In the case of double errors detected by the ECC, the complement retry feature is started.

There are three possible combinations of double errors: soft/soft, soft/hard, and hard/hard.

The first combination is not correctable. The second combination will be corrected. The third combination could be corrected. However, the correction is not implemented because of the risk that in subsequent memory accesses a soft

```
                                                            HH
1. Stored data            : 00000000 11111111 00000000 11111101
2. Fetched data           : 00000000 11111111 00000000 11111110
3  Store complctment      : 11111111 00000000 11111111 00000001
5. Fetched data           : 11111111 00000000 11111111 00000010
6. Complctment of fetched data: 00000000 11111111 00000000 11111101
```

Figure 59. Hard error correction by complement retry

error might combine with the two hard errors leading to an error not detected or misinterpreted. In the case of soft/soft or hard/hard errors an interrupt will be generated. In the following example (Figure 60) bit 30 is assumed to be a soft error, bit 31 is assumed a hard error. Further accesses to this memory address will only show the single hard error correctable by the ECC without activation of the complement retry feature. In the case of hard/hard errors both errors would have been disappeared after fetching of the complemented data, thus indicating the hard/hard error condition. In this case an interrupt would occur. In the case of soft/soft errors both errors would still be present after fetching of the complemented data so that no correction by the ECC can be performed. This indicates the soft/soft error condition and also causes an interrupt.

Figure 61 shows the timing of a retry procedure. Because of a double error, data D3 is not correctable by the ECC. DATA_INVALID indicates the faulty data bits on the STC bus. The STC resumes data transfer after 35 cycles with D3 corrected by retry.

2.5.3.4 Redundant Bit

Besides ECC and Complement Retry, the redundant bit feature improves data integrity.

Figure 62 shows a 16 MByte memory organization with 1 Mbit chips. In the example, the memory chips are x1-organized, which means they have a single-bit data-in/data-out. So, for a 4 Byte data bus, at least 39 memory chips are required (32 data bits, 7 check bits).

A single bit error in any of the memory chips will occur in only one of the addressable 4 MByte words (a word is equivalent to the width of the memory bus). A "chip-kill" however (a memory chip is faulty on all of its bits) will occur in all words to which this chip belongs (in this example 1 MByte words out of 4 MByte words). Therefore, an additional soft error will trigger retry and, even worse, a single additional hard error will result in an non-correctable error.

To avoid these conditions, the memory bus is not only 39 bits but 40 bits wide. The additional bit is called the REDUNDANT BIT (red bit). In the case of excessive hard errors on any of the other 39 bits the redbit can replace this bit

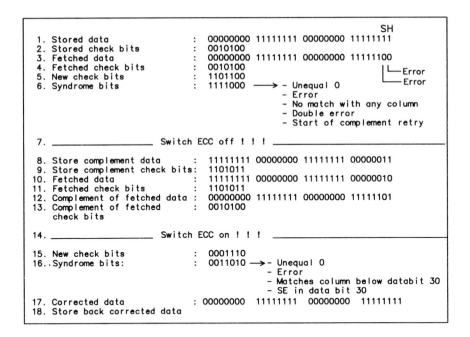

Figure 60. Double error detection / correction

position. For this purpose the STC chip includes a 16x6-bit register that holds the position (0-38) of the defect bit for each of the 16 banks (a bank is a memory part of 1 Mbyte).

Setting of this Redundant Bit Directory can be performed after memory test as part of the IML (Initial Microprogram Load) procedure, or "on the fly" after occurrence of the first soft/hard or hard/hard error.

In a store operation the bit to be stored on the position of the defect bit is also stored on the redundant bit. In a fetch operation the redundant bit (bit 39) is multiplexed on the position of the defect bit before error correction. The redundant bit can replace bank-selectively any of the 32 data bits or 7 check bits.

2.5.3.5 Scrub

Up to this point, all described error correction schemes are only activated if a regular read access to the specific address is performed. However, there are areas in the memory where, besides refresh, no regular access is performed for long periods of time. There is a higher risk of accumulating soft errors in these areas. The "Scrub function" reduces the risk that two soft errors on the same memory address result in an non-correctable error condition.

The Scrub function is controlled by the STC chip that "walks" through the entire memory. Every address is read and error correction is performed if necessary. After this, the data is stored back, hence preventing soft errors from

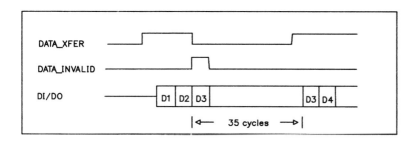

Figure 61. Retry timing

accumulating. For ease of implementation a Retry is performed on every address independent on the number of errors (no error, single error, double error). Scrub corrects data as well as check bits, and it is assured, that in case of an non-correctable error the data and check bits remain unchanged.

If the Scrub function detects an non-correctable error, no interrupt is generated because, for the running application, the data on this specific address may be unimportant or may be overwritten anyhow. The turnaround time of Scrub depends on the programmed refresh period and defaults to 1.14 h. Because of this long turnaround time scrub reduces the memory availability by less than 0.1%.

Scrub requests are issued by the refresh/scrub interval counter. The address to be scrubbed is provided by the Refresh/Scrub address counter (Figure 63) whose lower bits define the row address for refresh while all bits define the complete address (row, column, RAS) used for Scrub. If a Scrub request is issued the Scrub is appended to the next refresh. Execution of regular memory accesses is delayed like described for refresh (see "2.5.3.1 Refresh" on page 86).

Instead of this automatic Scrub, Scrub commands can be issued by the CPU chip. In this case, the automatic Scrub can be disabled via the control register, and CPU microcode has to manage the memory addresses and the Scrub turnaround time.

2.5.3.6 Address Fault Protection

All described features deal with data integrity such that in the case of a data error, recovery is performed. They are used to correct failing bits of the memory. Mis-addressing, however, (an error condition that reads or overwrites data by accessing to wrong addresses) can not be detected by these features.

A memory is addressed via multiple signals: RAS, CAS, DOG, DL, and Addresses. All these signals, except Addresses, have an active and a passive level. Only in the case of an active level will they trigger operations of the memory. Hence, if they do not work properly, e.g. if they are tied to ground by a defect in the driver logic, the error will be detected due to the malfunction of the memory.

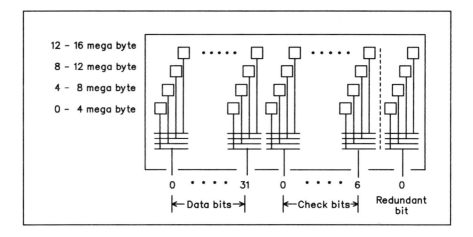

Figure 62. Memory organization

A defect in the Address lines, however, will not be detected. High level as well as low level are valid and will address the memory. Thus a stuck fault on an address line will cause other addresses than intended to be accessed. To detect these address faults the overhead in the check bit matrix (see " 2.5.3.2 ECC (Error Correction Codes)" on page 88) is utilized. Two more columns are added to the matrix, and in addition to the 32 data bits two parity bits, generated out of the address bits, are included in the check bit generation. Because of the row/column address multiplexing, two separate parity bits for row and column address have to be used. In this way not only information on the data but also on the address is stored in the check bits. If, due to a fault in the Address lines, a wrong memory address is accessed, the ECC will detect an error. This error will appear as an non-correctable soft/soft error, triggering an interrupt.

2.5.4 Diagnostics

In addition to its regular modes the STC chip can run in specific test modes to facilitate defective diagnostic of the STC and the memory. The test modes are run either after power-on as part of the IML procedure to check the memory for errors and set the Redundant Bit Directory, or in case of an error condition to localize the error. The STC is set to these test modes by bits in the control register (see "2.5.1.2 Error Correction Unit" on page 79). The following is a list of some of the test modes:

INHIBIT ECC disables error detection in fetch mode although check bits are generated in store mode. This feature is used to check the memory for errors. In DISABLE RETRY mode, only single errors are corrected by the ECC. Any double error will trigger an interrupt.

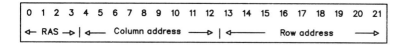

Figure 63. Refresh / scrub address counter

With TEST ECC activated, every store command writes "0" to the memory on the data bit positions, while the check bits are generated according to the data bits of the store command. Hence soft errors on any data bit position can be generated. In FORCE ERROR mode, bit 0 of the memory bus will be forced to "1" during fetch operations, simulating a hard error. The use of TEST ECC and FORCE ERROR allows the STC chip to be checked for correct detection/correction of single and double errors as well as soft and hard errors.

During CHECK BIT READ OUT, the check bits will be placed on the STC bus in fetch operations. Like INHIBIT ECC, this feature is used to check the memory for errors. To test the STC parity check, the FORCE PARITY ERROR mode inserts a parity error on the STC bus. Five bits in the control register define the byte and whether the parity error shall occur in command/address or data transmission.

2.5.5 Personalization

After power-on the STC chip and memory are initialised in several steps (see Figure 64). The initialization starts with setting the STC latches to their default values by shift cycles of the LSSD chains (chain flushing). Deactivation of a clock-stop signal triggers the STC to enable the off chip drivers, and to start refreshing the memory.

Memory identification is performed by CPU microcode to obtain information on memory card type, memory organization, and memory size. As a result of the identification procedure the configuration, memory card type and memory size will be stored in the STC internal control registers. For identification purposes the memory card identification bits (ID bits) have to be read. ID bit reading is achieved by setting bits in the control register, so that subsequent fetch commands will place the ID bits of the addressed card on the STC Bus.

Memory test and redbit setting are also performed by CPU microcode. One defect bit position out of 39 (32 data bits, 7 check bits) per bank can be replaced by a redbit. The information on the position of defect bits is stored in the redbit directory. Finally, memory initialization is performed to store correct check bits into the memory.

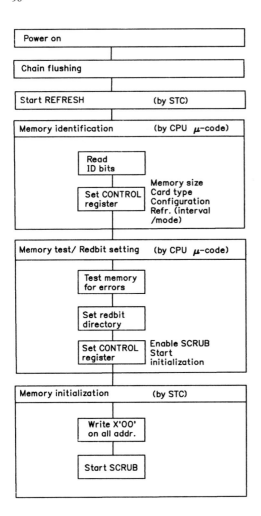

Figure 64. Start-up procedure

2.6 Floating Point Coprocessor

Jean-Luc Peter, Claude Sitbon, Wolfgang Kumpf

Existing computer architectures employ a variety of different, and incompatible, representations of floating point numbers. The S/370 architecture defines a hexadecimal floating-point format. A part of the industry is moving to the IEEE 754 binary floating-point format, which, due to its normalisation characteristics, is slightly more precise than the S/370 hexadecimal format. The IEEE 754 standard is still evolving, and lacks for example an equivalent to the S/370 extended floating-point number format. Different from the VAX architecture,

Floating Point Coprocessor

which has implemented a dual floating-point number standard, the S/370 architecture has decided to stay with the existing hexadecimal standard.

There are several reasons for it. The user community has made over the years a huge investment in recording and storing large quantities of S/370 floating-point data. The difference in precision between S/370 and IEEE 754 floating-point numbers is small and often does not justify the effort to convert existing data. The recent introduction of guaranteed accuracy arithmetic (ACRITH) on all S/370 systems makes the discussion of precision in number representations less relevant [IBM2]. Perhaps more important, many experts feel the future may belong to a yet to be defined decimal floating-point standard, and that the presently evolving binary IEEE 754 standard may have only temporary significance. A possible future decimal standard is expected to eliminate many rounding and number conversion problems inherent in the existing floating-point number representations.

2.6.1 General Description

The Floating-Point Unit (FPU) chip was designed by the IBM Essonnes Component Development Laboratory. It executes the complete set of S/370 Floating-Point Instructions in short (32 bit), long (64 bit) or extended (128 bit) format representing a total of 51 instructions. In addition the chip executes a subset of 9 ACRITH (High Accuracy Arithmetic) instructions, 2 square root instructions and 6 special instructions for status handling and data exchange.

The /370 architecture specifies 4 floating point registers of 64 bit length each. These registers are included on the FPU chip. Since the Capitol chip set will work without a floating point coprocessor chip, performing floating point operations in microcode, a duplicate of the floating point registers is included on the CPU chip. These registers remain unused when an FPU chip is attached.

The chip is optimized for attachment to the Processor Bus. For floating-point operations the 32-bit wide instruction/data bus is shared by the CPU, MMU and FPU chips and allows fast data exchanges between the FPU and the main store.

Since floating-point operations require a very high reliability related to data integrity, the FPU function has to be supervised by complex check mechanisms. In the Capitol chip set two identical FPU chips, connected in parallel, are continuously executing the same task on a cycle by cycle base and in each cycle the outputs of both chips are compared and checked for identity ("Shadow" approach).

2.6.2 Floating-Point Instructions and Data Format

In general floating-point instructions are used to perform calculations on operands with a wide range of magnitude and to yield results scaled to preserve precision.

The S/370 architecture provides floating-point instructions for loading, rounding, adding, subtracting, comparing, multiplying, dividing, and storing as well as controlling the sign of short, long, and extended operands. Short operands generally permit faster processing and require less storage than long or extended operands. On the other hand, long and extended operands permit greater precision in computation. Floating-point instructions may be performed either as register-to-register (RR) or as storage-and-register (RX) operations.

In the IBM S/370 architecture a floating-point number is expressed as a fraction multiplied by a separate power of 16 (hexadecimal representation). The term floating-point indicates that the radix-point placement is automatically maintained by the machine.

The part of a floating-point number which represents the significant digits of the number is called the fraction. A second part specifies the power (exponent) to which the base is raised and indicates the radix-point of the number. An example illustrating these terms is shown in Figure 65. by an example. Please notice that the IBM S/370 architecture defines the term characteristic which represents the signed exponent. The characteristic is obtained by adding 64 to the exponent value (excess-64 notation). The range of the characteristic is 0 to 127, which corresponds to an exponent range of -64 to +63. In this way the exponent value is signed in an easy way. Figure 66 shows the three possible S/370 floating-point formats. The fraction and the characteristic may be represented by 32 bits (short format), 64 bits (long format), or 128 bits (extended format). A floating-point number has two signs: one for the fraction and one for the exponent. The fraction sign, which is also the sign for the entire number, is the leftmost bit of each format (0 for plus, 1 for minus). The sign for the exponent is obtained by expressing the exponent in excess-64 notation with the term characteristic as already mentioned above.

2.6.3 FPU Interface and Communication

In Figure 67 the configuration of the involved chips for a floating-point operation is shown. The two FPUs are connected with the Processing Unit (CPU) and the Memory Management Unit (MMU) by the Processor Bus. Both FPU's, one for the floating-point task and the other for the "shadow approach" function, are connected together by a special bus for the compare signals to detect possible FPU malfunctions. In addition there are some lines, bypassing the Processor Bus, for signals dedicated to CPU-FPU communication (e.g. Busy/Idle signal, Condition Code signals, FPU interrupt signals caused by exceptions, exponent underflow/overflow etc.).

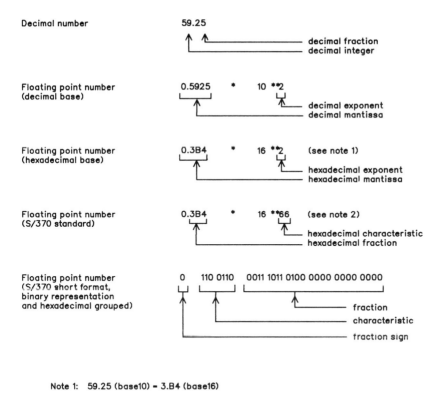

Figure 65. Different number representations by conversion (example)

The FPU chip acts as a slave of the CPU chip and cannot initiate any communication. Only the CPU chip processes the S/370 instruction stream and executes all instructions except floating-point operations. If a valid S/370 Floating-Point instruction was recognized the CPU chip generates a message which is sent to the FPU chip (1). The format of this message is defined by the Processor Bus protocol and contains the S/370 Floating-Point OP-Code, the "From-To" unit addresses and the affected Floating-Point registers located in the FPU.

In case of RR-instructions two Floating-Point register addresses containing both operands have to be transferred and the instruction can be performed by the FPU chip immediately after receiving the message (2-party communication). When the instruction is executed this is signalled to the CPU chip by the BUSY/IDLE signal and the CPU chip may proceed with the next instruction.

If the CPU chip has decoded a RX-instruction, only one Floating-Point register address for the first operand can be transferred. The second operand is located in main storage (or the Cache) and the CPU chip performs the address calcu-

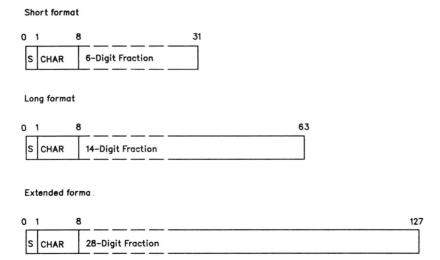

Figure 66. S/370 floating-point number representation

lation to initiate the data fetch by sending a message to the MMU (2). The MMU then picks the data out from main storage or Cache to put it on the Processor Bus (3), where it is picked up by the FPU. This kind of information exchange is called 3-party communication. For RX instructions with long format the data fetch actions (2) and (3) have to be performed twice. After receiving correct data the FPU chip starts execution of this Floating-Point instruction. During that time the CPU chip watches the Busy/Idle signal and proceeds after the FPU chip has finished the processing.

By decoding a S/370 Instruction of the type "Store" the content of a Floating-Point register has to be written into main storage. The CPU chip again performs the address calculation, sends a message to the MMU (2) and the FPU sends the correct data via Processor Bus and MMU chip to the main storage in the next cycle (4).

FPU chip (B) watches all data sent from the functional FPU chip (A) and compares with its own data generated in parallel. All drivers of the FPU shadow (B) are degated, only the chip status with the check information may be read out to analyse error conditions in case of mismatches between FPU chip (A) and FPU chip (B).

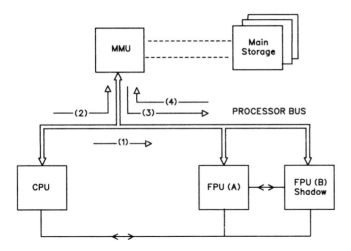

Figure 67. The FPU in the CAPITOL chip set environment

2.6.4 Chip Logical Description

2.6.4.1 Overview

The FPU chip contains five logical units:

The hardwired **Control Logic** for the instruction decode and for driving the various data flow parts and internal storage units.

The **Internal Dataflow** for fetching the operands, executing the FPU instructions and storing intermediate and final results. It includes a 9-bit wide data flow for exponent calculation, a 62-bit wide data flow for fraction (mantissa) calculation and the four architectured Floating-Point Registers for storing operands and final results. The Floating-Point Registers are logical splitted into two parts fitted to the independently working exponent and fraction units.

The **Bus Interface** logic for receiving instructions, data and status information and sending data and status information in line with the Processor Bus protocol.

The **Checking** logic for Error Detection and Fault Isolation (EDFI), to record the various types of malfunction and to stop the machine in case of checks.

The **Unit Support Interface (USI)** logic for service processing, see "2.11.6 Support Interface" on page 141.

In Figure 68 the main components of the FPU internal data flow are shown. It provides a 64-bit data width matching to S/370 floating-point instructions with long format. Since the Processor Bus width is 32 bits there are two Processor Bus cycles required for the long format data transmission. The floating-point exponent and fraction processing is executed by two independent parts of the chip dataflow working in parallel as a rule.

2.6.4.2 Exponent Dataflow

The exponent dataflow handles the 7-bit exponent data extended by two special bits for detecting and saving of exponent-overflow and exponent-underflow conditions. The exponents are processed in 64-excess notation in identical fashion for short, long or extended format operands. The related dataflow units are a 9-bit adder, the 4x8-bit Floating-Point register part for the exponent including the sign-bit and two internal work register.

The Exponent Adder Unit performs add, subtract, compare, increment, and decrement exponent operations. It may also be used as a loop counter for the multiply, divide and square root algorithms. This adder unit is used for all arithmetic instructions and takes one machine cycle for each operation.

2.6.4.3 Mantissa Dataflow

The mantissa dataflow is 62-bit wide and handles the 14-digit fraction data (matching the floating point long format) extended by the Guard digit, the Rounding bit and the Sticky bit. These six additional bits (Guard digit, Rounding bit and sticky bit) are necessary for correct rounding of intermediate-results by executing ACRITH instructions. To perform the different rounding operations, most floating-point instructions with rounding options must temporarily retain additional intermediate-result information to the right of the 14-digit (6-digit) wide fraction of the long (short) format. Whenever an operand fraction or an intermediate-result fraction has to be shifted right, digits shifted out of the rightmost fraction digit enter the Guard digit position, digits shifted out of the Guard digit enter the Rounding bit position, and the bits shifted out of the Rounding bit are logically ORed into the Sticky bit. During any arithmetic operation this relevant information participates in further data processing and may have an effect upon the final result, depending of the operand values and the rounding mode.

The FPU mantissa dataflow consists of the following major units:

The Fraction Adder, realized as a Carry Select Adder, performs Addition and Subtraction of the 62-bit wide extended operand fractions, provides results with absolute or relative values, and is used for nearly all arithmetic instruction types. A basic 62-bit add operation requires two machine cycles except for the Divide instruction. Since this instruction is executed with a 1-bit algorithm, the adder function for the divide instruction was optimized to be performed in one cycle only.

The Shift Unit, implemented as a three stage barrel shifter, allows shift right and shift left up to 15 digits within a single cycle. It also can force bits to logical zero or one and is used in almost all arithmetic instruction types.

Floating Point Coprocessor

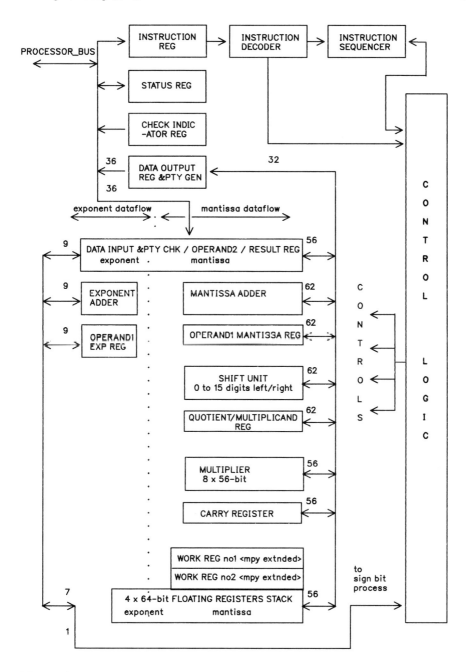

Figure 68. FPU chip global data flow

The Multiplier Unit is a hardwired 8 x 56-bit array multiplier. It is based on the Booth encoding algorithm and delivers an 8-bit result per cycle. This part of the hardware is used in Multiply instruction types only.

The four Floating Point Registers, defined by the S/370 Floating-Point architecture, are an on-chip 4x64-bit stack of non-LSSD registers storing the sign, the exponent part and the fraction part of operands and results. This stack also includes two additional 56-bit work registers to store intermediate data dedicated for internal use. One read or write access initiated by the CPU is performed in one cycle.

2.6.5 Reliability, Checking and Testing

2.6.5.1 Reliability

Different from most other parts of a processor, silicon problems in the data flow area of an FPU function are often difficult to detect. These errors typically occur as a kind of result "deviation". The grade of the inaccuracy depends on the location of the error within the data flow, and on the instruction itself.

Plausibility checks in the user program task may detect these wrong results, but still may interpret them incorrectly as a problem of accuracy instead as a hardware problem. It is very important to have good check mechanisms or algorithms in the floating-point area to guarantee the accuracy of the computed results.

2.6.5.2 Checking

The FPU chip does not employ the sophisticated approach to checking and testing found in the other components of the Capitol chip set. Instead, this FPU function is performed by two identical chips that are connected in parallel and are continuously executing the same task on a cycle by cycle basis. At each cycle the outputs of both chips are compared and checked for identity (shadow approach).

This duplicated hardware approach allows to eliminate special features like parity predict or modulo 3 checking schemes inside the data flow structure. This silicon gain can now be used to focus on larger/faster executions units without the usual penalties on critical paths introduced by complex checking schemes.

The FPU chips (A) and (B) are identical parts but are wired differently on the card to achieve their specific function. Each chip implements byte parity checking/generation of the Processor Bus at the input/output pins. In addition every instruction fetched into the FPU chips is internally checked for valid code combination (e.g. S/370 OP-code, "From-To" unit addresses, Floating-Point register addresses etc.).

Both chips are executing a given task independently and the "Shadow" chip is comparing intermediate and final results at the outputs and checks them for identity in each cycle.

Any error occurring during task processing is recorded in dedicated registers on the FPU chip, that may be sensed via special CPU microinstructions. The set of the on-chip checkers necessary to the "Shadow" approach, the input parity bits, and the valid instruction checkings can be individually tested for correct function via additional special FPU microinstructions.

The extra cost of the "Shadow" approach (one extra module) has been considered "acceptable" taking into account:

> The capability to detect approx. 98% of any single or multiple "stuck-at" faults on the next machine cycle following error occurrence.

> The low silicon overhead within each chip: only XOR ckts are required to compare outputs of both chips.

> The low complexity of the logical design (compared to the standard checking schemes)

2.6.6 Performance

The following are the FPU chip execution times for some typical floating-point instructions assuming an 80 ns clock cycle time:

S/370 Format:	Short (32 bits)	Long (64 bits)	Extended (128 bits)
ADD/SUB	.56 μsec.	.56 μsec.	1.52 μsec.
MULTIPLY	.88 μsec.	1.28 μsec.	4.96 μsec.
DIVIDE	2.88 μsec.	5.44 μsec.	---

Since the execution times are data dependent, actual timings are often faster (in case of the ADD/SUB instructions about 60%, in case of the MULTIPLY instruction about 85% of the values shown). Please note that the overhead required to build and send a message via the Processor Bus and also the time for fetching storage operands is not included.

A system environment with the FPU chips imbedded in the Capitol chip set executing the WHETSTONE benchmark in short and long precision with an 80 ns cycle results in 1.4 MIPS with long precision, and 1.8 MIPS with short precision. The corresponding LINPACK numbers are 0.36 MFLOPS with long precision and 0.45 MFLOPS with short precision.

2.7 Bus Interface Chips

2.7.1 Overview

I/O Devices like Disk, Display, Communications and others are connected to a microprocessor via I/O adapters. It is possible to connect I/O adapters directly to the Processor bus, as shown in Figure 69. For example, the IBM Personal Computer or low end VAX systems have such a structure.

For better I/O performance a structure as shown in Figure 70 is universally used. Here, the Processor Bus is interconnected via a Bus-to-Bus Adapter (BBA) to an I/O Bus that attaches all I/O adapters. The Capitol chip set Processor Bus design assumes this approach.

Figure 71 shows the Capitol chip set as configured in the IBM ES/370 model 30. (The implementations in the IBM ES/370 model 50, and the IBM 3092 Processor Controller models 004 and 005 are very similar.) Here the processor bus interconnects to 2 I/O busses to increase I/O performance. Two VLSI chip types are employed, the MBA and the BCU chips, that implement the Bus-to-Bus adapter (BBA) function. The IBM ES/370 model 50 looks similar, except that a single MBA chip connects to 4 BCU chips and 4 I/O busses. Every BCU chip has access to the complete main store, including the non-S/370 memory. The BCU chip performs address and key checking and the MMU cache is accessed, if the data resides in the cache. This mode of operation is used for Direct Memory Access (DMA) transfers initiated by an I/O adapter, attached to an I/O bus, and may occur during data transfer, or the fetching of S/370 I/O commands (channel command words, CCWs).

The Capitol chip set Processor Bus is capable to transmit 4 bytes every machine cycle. With an 80 ns cycle this implies a maximum Processor Bus data rate of 50 MByte/sec (operational data rates are somewhat lower). The ES/9370 I/O bus data rate is approx. 5.5 MByte/sec.

The BCU chip is used in other ES/9370 systems as well, e.g. the ES/9370 Model 90. The MBA chip design is unique to the Capitol chip set, and adapts to the existing BCU interface. Because of minor variations in the MBA to BCU interface between various systems using the Capitol chip set, 4 sightly different versions of the MBA chip had to be designed.

It is important to differentiate the hardware configuration of Figure 71 from the I/O configuration as seen by the architecture. Here, the attachment of a large number of S/370 channels to a single I/O bus is possible. E.g., multiple disk or tape I/O adapters may attach to a single I/O bus, that may simultaneously perform multiple I/O operations (execute multiple S/370 channel programs). However, due to a bandwidth limitation of 5.5 MByte/s, only a single 3 MByte/sec disk devices like the IBM 9332 or 9335, may transmit at the same time over an I/O bus. Thus, 2 I/O busses are required to permit simultaneous data transfer of 2 high speed disk devices.

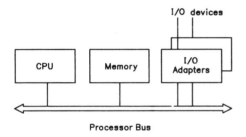

Figure 69. Single stage bus structure

2.7.2 MBA Chip

As indicated in "2.10.2 Processor Bus Implementation" on page 128, the MBA chip may serve as the Processor Bus master, initiating commands and the transfer of data. Except for DMA operations, information transfer between main storage and the BCU chips is done through a predefined area in non-/370 memory. This "hardware I/O buffer area" (unrelated to an I/O buffer as an assembler or HLL program sees it) is shown in Figure 72. Its base address and the length of the message queue area are stored in a register on the MBA chip. Each BCU chip owns a 64 Byte mail box area in the hardware I/O buffer area.

The MBA chip contains registers that store the required pointers to access main store directly. For speed matching reasons, it also contains an on chip 64 Byte internal buffer for each attached BCU chip, into which the data are being loaded. On Write operations, the MBA chip prefetches data from main memory.

Let us look at an I/O data transfer operation as performed by a S/370 Write Channel Command Word (CCW). The I370 microcode interpretation routine that executes the CCW first calculates the I/O buffer address in the S/370 address space. It then moves a message into the mailbox associated with the appropriate BCU chip. A special control microinstruction sets a latch on the MBA chip, that signals to the BCU chip the availability of data. The BCU chip instructs the MBA chip to fetch the content of its mailbox. The BCU chip interprets the message, and recognizes a DMA request. It communicates over the I/O bus with the appropriate I/O adapter to request the data. Simultaneously it sets up the pointer registers in the MBA chip. When the data arrive over the I/O bus, the BCU chip loads them into the 64 Byte internal buffer on the MBA chip (the MBA chip has a separate internal buffer for each attached BCU chip). The internal buffer is a speed matching device. The Processor Bus is able to accept data much faster (max. 50 MByte/sec), than the I/O bus can deliver them (approx. 5.5 MByte/sec.). The pointer registers on the MBA chip assure loading of the internal buffer content into the main memory at the address specified by the Channel Command Word. Address incrementing and length count is handled by the MBA chip.

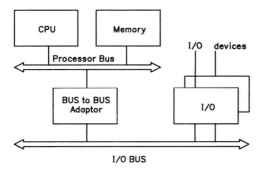

Figure 70. Dual stage bus structure

A read operation works in reverse in a similar fashion. I/O adapters attached to the I/O bus can generate CPU interrupts by sending messages via the BCU and MBA chips into the inbound message queue area of Figure 72. Multiple messages originating from different I/O busses are queued in chronological order in this area. Message are 16 Byte long. Appearance of a message causes a CPU interrupt, provided the CPU chip is ready to accept it. Two pointer registers in the MBA chip manage the first and last entries in the menage queue.

2.7.3 BCU Chip

The IBM 9370 uses a normal asynchronous I/O bus with address, data, command, and status transfer capabilities. The BCU chip and the equivalent functions on the I/O adapters manage communication in the usual fashion. The BCU chip controls I/O bus arbitration and monitors all ongoing bus operations with the help of a timer. It performs Direct Memory Access (DMA) to main memory, and handles message acceptance and message origination. On a timeout it collects status information and requests support action by the CPU chip. There are relaxed timing requirements due to the buffer in the MBA chip.

The BCU chip has access to the complete main memory, including non-S/370 memory. It is not timed by the clock chip, and has an asynchronous interface to the internal buffers on the MBA chip. On a "Write" operation the internal buffer on the MBA chip collects the data transmitted from the BCU chip, and sends it to the Processor Bus. On "Read" operations, the MBA chip prefetches the data from main memory and delivers them to the BCU chip on a word basis at the required speed.

Bus Interface Chips

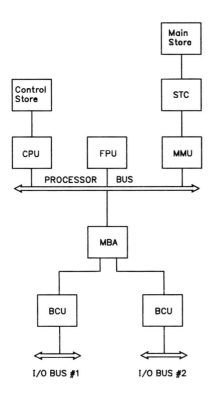

Figure 71. I/O bus interconnection

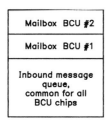

Figure 72. Hardware I/O buffer area in non-/370 mainstorage

2.8 Clock Chip

Klaus-Dieter Müller, Dietmar Schmunkamp

2.8.1 Central Clock Generation Versus Distributed Clock Generation

The machine cycle is the basic unit of timing in a computer. Within a machine cycle atomic operations, e.g. performing an addition, take place. Instructions are executed within one or several machine cycles. The machine cycle is controlled by an external timing source, an oscillator (usually a crystal) with a constant frequency. Clock logic uses this input to generate various timing signals to control the processor logic at desired timing points within the machine cycle.

Existing microprocessor designs implement the clock logic function in one of two different ways. The Motorola 68020/68851/68881 family integrates the clock functions into the individual CPU, MMU, Floating point coprocessor etc. chips. The Intel 80xxx family utilizes a separate clock chip.

The Capitol chip set follows the second approach: A separate clock chip generates the clock signals for all other chips. As those portions of the system that run synchronously usually span several chips, the design of the clocking system deserves special attention. If clock signals are distributed over several chips the individual chips may be drastically different with regard to process tolerances. Thus the same clock signal may arrive at the using logic spread out in time, referred to as clock skew. A centralized clock chip, together with a carefully designed topology for clock signal wiring, significantly reduces clock skew. Effects of clock skew are discussed in "2.9.4 Evaluation of the Clock Skews" on page 120.

In addition to improved skew control, central clock generation permits to perform error detection and error recovery on the system level rather than on a single chip basis. For this the clock-chip provides some basic run controls such as start-stop and single cycle.

The clock chip requires two external components: a hybrid oscillator, and a delay line. This configuration is shown in Figure 73. The clock chip produces multiple clock signal outputs. It selects for each output signal slope a timing point which is predefined by the tabs of the delay line through which the hybrid oscillator is feeding the clock generators.

Using an active delay line eases the card test and ensures superior noise immunity and accuracy. In addition to the clock generation the clock chip performs functions like Start - Stop control, Power on reset recognition, Reset sequence generation for other chips, Clock checking, and Console key controls.

Clock Chip

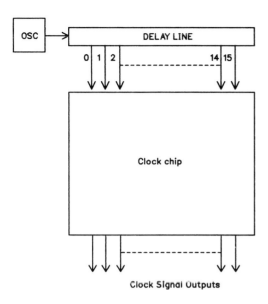

Figure 73. Clock chip configuration

2.8.2 Logical Implementation

An overview of the Clock chip logical structure is shown in Figure 74. Clock signal generation is implemented with the help of 52 clock generators on the chip. There is a "clock macro", which is a circuit configuration repeated 52 times on the chip. The clock macro itself is shown in more detail in Figure 75.

The tabs of the external delay line are connected to 2 groups of 8 receivers circuits. Typically, the leading edge of the clock signal is derived from one group, while the trailing edge is derived from the other one. Eight of the 16 receiver circuits feed through buffers and unidirectional busses 52 selectors A which are part of the 52 clockmacros. The other group of eight receiver circuits feeds 52 selectors B. This is shown in Figure 77. Each selector picks up one of its 8 input signals to clock a 6 bit shift register (Figure 75) with feed-back from bit 2 to bit 5 or from bit 0 to bit 5 (programmable).

(The clocking signals are conditioned by a clock splitter, CLSP, which generates non-overlapping LSSD-clocks for the shift register. The LSSD approach is described in "5.1.5 LSSD (Level Sensitive Scan Design)" on page 248).

Bit 5 of shift reg A and bit 5 of shift reg B are combined by a circuit which performs either an AND or an OR operation depending on the state of its control input. The shift register A generates one slope of the output signal and shift register B the other one. The output of the combiner is directly connected to 2 CMOS drivers.

For maximum flexibility, the clock chip has been made programmable. Each clock generator is independently programmable as far as pulse width, phase

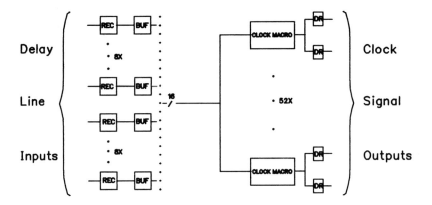

Figure 74. Clock chip overview

(and cycle time) is concerned and thus can easily be adapted to the varying system requirements. The 8 control lines within the clock macro originate from an 8 bit program storage built by LSSD latches. For programming the clock macro, information has to be stored in the program storage latches and in the shift register latches.

Figure 76 shows how the timing of one clock pulse is derived from a delay line tab A (OSC) and a tab B (DOSC).

The clock chip is initialized by loading the 8 bit program storage and both shift registers in each of the 52 clock chip macros. The clock initialization pattern is obtained from one of 3 hardwired patterns or an external source (shift chain or EPROM). Selection of the initialization source can be programmed via two chip pins. The initialization is triggered by power-on recognition, as a result of a clock check, or via the Support Bus. Using an EPROM requires an external 9 bit counter. The clock and reset signals for the counter are generated by the clock generation chip.

There are other alternatives to load the program storage on the clock chip. Like the other chips in the Capitol chip set, the clock chip has an Unit Support Interface to the Support Bus (see "2.11.6 Support Interface" on page 141), that may be used for this purpose. In addition, there are 3 default timings, that can be activated by applying the correct signals to 2 external pins.

2.8.3 Timing Tolerances, Reset, and Checking

Each clock macro output is able to drive a capacitive load of up to 20 pF with the specified tolerances. As there are two clock output driver circuits for each clock macro, a total capacitance of 40 pF can be driven. If this is not sufficient several identically programmed clock macros can be used to increase the driving capability. In this case the clock lines must be dotted at the receiving chip pin. All clock lines are in general point to point connections. The clock

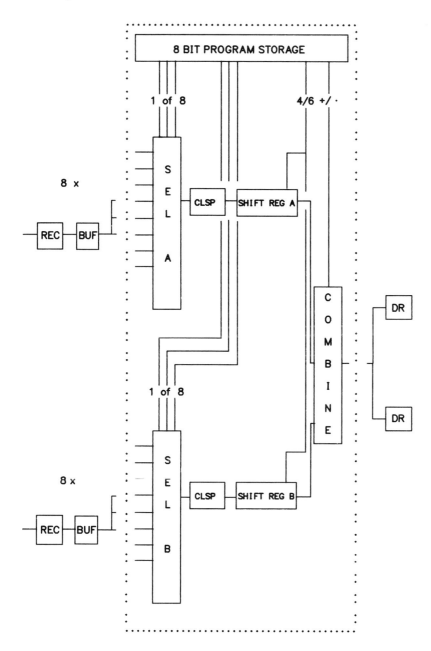

Figure 75. Clock macro overview

drivers use the full swing of the power supply voltage to achieve good noise immunity on the clock lines.

The Clock chip is the master of the chip set for the reset function. The other chips in the chip set require different resets according to S/370 architecture and

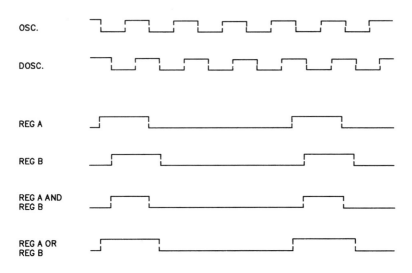

Figure 76. Clock pulse timing

implementation requirements. Resetting the Capitol chip set is done by loading zeros into the SRL (Shift Register latches) described in "2.11.6 Support Interface" on page 141. For this purpose the SRL latches are divided into 3 groups: a) function (F-)latches, b) check indication (I-)latches, and c) Support Bus interface (U-)latches. The Clock chip performs the reset for the other chips in the chip set by rippling a forced zero level serially through a SRL chain in 2048 A/B shift cycles. Arrays are reset by microprogram. There are 5 reset functions:

"Power On Reset" clears and initializes the Clock chip first. Then the other chips of the chip set are reset under clock control by resetting the F-, I- and U-latches. "Reset for IML" resets the F- and I-latches in the other chips of the chip set. "System Reset" controls, according to S/370 architecture requirements, the reset functions required for Normal, Clear, Load, and System. It also resets the F- and I-latches in the other chips of the chip set. "Check Reset" is initiated if one of the 4 STOP lines is activated within the chip set. The Clock chip stops the distribution of the functional clock signals, thus causing a halt of the function execution. A 2048 msec wait time is introduced to allow intermittent failures to settle. Subsequently 2048 cycles are allocated for resetting the F-latches in the other chips. Finally a 2048 cycle wait time is allotted for memory synchronization and adjustment of the speed control for the chip drivers in the chip set.

"Start After Reset" supplies the functional clocks again. Any reset condition also forces a reset condition within the CPU chip. This starts an initial 64 Byte microprogram fetch from the EPROM address space and passes on control to the beginning of that microprogram block (see "2.2.3.2 Micromode" on page 25). The microprogram proceeds depending upon the reset type which

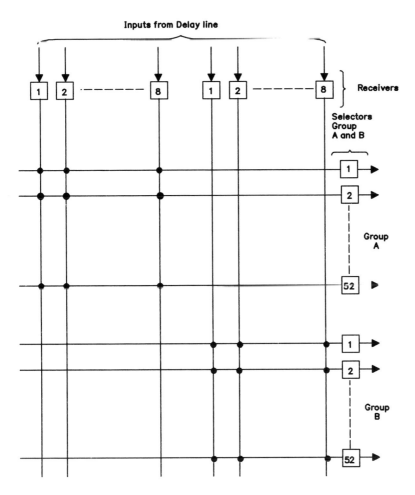

Figure 77. Distribution of delay line inputs to selectors

has been passed on from the Clock chip to the CPU chip via 3 dedicated reset control lines.

As the clock generation depends on the contents of latches which may change erroneously by an external event, a clock checking mechanism is implemented on the chip. This guarantees validity of the clock signals.

The clock check mechanism uses a parity check on all latches which are involved in clock generation. The parity of these latches is compared with a predicted value. In case of an error the latches are reset to their initial state and a clock check indication is given to the system.

The Clock chip provides an interface to the Support Bus. An external device can be used to access and manipulate internally latched data on the Clock chip, provided the clock is stopped.

2.9 Clocking

Hermann Schulze-Schoelling

2.9.1 Clock Signal Types

The various clock signals (clocks) available at the output of the Clock chip are distributed to the other chips in the Capitol chip set. Each logic chip uses a basic set of clocks:

A1-clock	shift clock	continuously running
B1-clock	slave latch clock	continuously running
CC-clock	master latch clock	continuously running
C1-clock	master latch clock	stopable clock
D1-clock	bus driver enable	stopable clock

Additional clocks are distributed to the chips for the array macros, control store array, mainstore interface timing, and I/O bus interface timing. Figure 78 shows an overview of all clock signals distributed from the Clock chip to the logic chips together with their respective number of parallel clock drivers.

2.9.2 Clock Generation Flow

The Clock generation flow is shown in Figure 79. The clock chip uses three-state output drivers that can be forced into a high impedance (HZ) state. This function is used to "stop" the clock by forcing the driver into its HZ state while the clock is already in the off-state. The respective off-state level will be held by a 10 KOhm resistor, which is either connected to ground or 5V. The "start" of the clock is performed by the deactivation of the HZ state as long as the clock is in its off-state. This method saves one AND-gate in the clock generation path and reduces the stop gate implementation effort.

The clock signals are distributed on the card with an adjusted length of 5 inch. The clock driver may drive a maximum capacitive load of 20pF. This is to guarantee a certain waveform of the clock pulse. If the capacitive load exceeds the limit of 20pF, additional clock drivers are connected in parallel to the net. They are dotted at the receiving module pin. The clock timing specification reflects this point in time when the clock pulse arrives at the module pin.

The clock pulse is received at the logic chip by a CMOS receiver and distributed on the chip to the respective latches.

Clocking

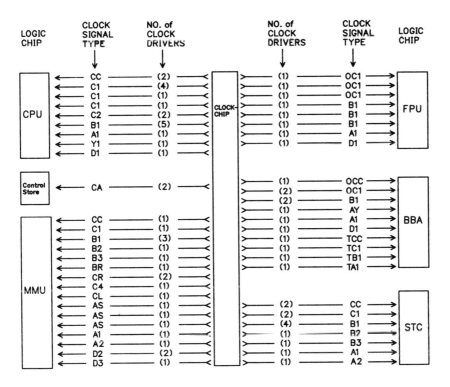

Figure 78. Clock signals overview

2.9.3 Clock Pulse Generation

The clock macro contains a program store as discussed in "2.8.2 Logical Implementation" on page 113. Its information content is used to generate the appropriate clock pulses from the oscillator pulses. The 16 oscillator pulses are separated in two groups, each of which contains 8 pulses. In general, one oscillator pulse of any group is selected to set a clock chip signal output latch (clock latch). A later oscillator pulse, but from the other group, is selected to reset the latch and deactivate the clock signal. The precision of this function is guaranteed by using always the down going transition of the respective delayed oscillator pulse to perform either the set- or the reset of the clock latch.

Three oscillator periods of 26.66 ns (37.5 MHz) form a total cycle of 80 ns (see Figure 80).

Most of the clock signals repeat in each machine cycle. However, there are several clocks with a repetition rate of 2:3 that means, there are three clock pulses within 2 cycles. This results in a cycle time of 53.3 ns, which is used by the BCU chip (see "2.7 Bus Interface Chips" on page 108). Furthermore the clock macro logic distinguishes between a negative or positive active clock pulse. This occurs at the output of the combiner circuit (see Figure 75).

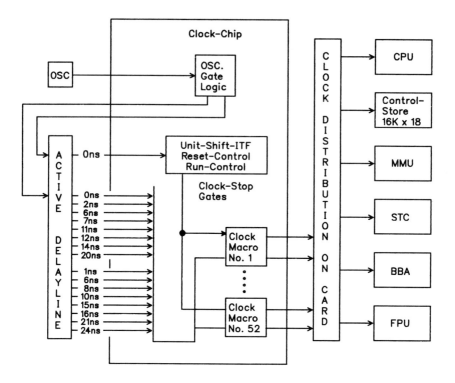

Figure 79. Clock generation flow

2.9.4 Evaluation of the Clock Skews

Each stage or partition along the clock generation path may add a certain clock skew to the clock pulses. This means the clock pulse at the destination (latch or array) may have a certain offset relative to another clock pulse. Any additional circuitry, employed to keep this within limits, has to assure that the effort of testing the card with its VLSI chips remains reasonably low. Also the large VLSI chips imply a significant number of latches. The CPU chip for example has more than 1600 shift register latches (SRL's). The average capacitive load of the latch is 250fF, which is the sum of the wiring and the fan-in capacitance. The total capacitance is 400pF, which is a large value for the clock powering.

The entire clock generation path contains four parts which contribute to the clock skew. They are the delay line, the clock chip itself, the different loads of the clock drivers, and the skew between logic chips.

The "delayline dependent clock skew" is generated in the delayline logic and the delayline drivers. The vendor specification allows a maximum skew of 1 ns / delayline tap in reference to the zero-delay tap, e.g. if a particular pin is specified at 10 ns delay, the actual delay may be between 9.5 ns and 10.5 ns. This variation does not track between delayline taps.

Clocking

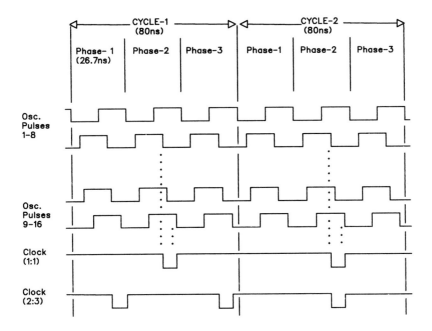

Figure 80. Clock pulse generation

The "capacitive load" of the clock chip output signal (the sum of module and chip input capacitances) may differ due to process tolerances and differences in the clock powering scheme on the receiving chips. The variation ranges from 5pF to 20pF / clock driver. If the load is greater than 20pF multiple clock chip outputs are connected in parallel. The skew is within 1 ns between two different clock signals (a lightly loaded clock may arrive 1 ns earlier at the receiving module pin than a heavy loaded clock signal). The skew due to differences in wiring length can be neglected. Assuming the wiring length is nearly identical, the corresponding skew is a small fraction of 1 ns.

The "clock chip dependent skew" is produced in the clock chip itself due to process tolerances (tracking). The internal path delays may also vary due to slightly different wiring length in the clock generation paths. Otherwise identically timed output pins may feature a skew of 3 ns. This means, any clock transition to any other clock transition may have a skew of 1.5 ns.

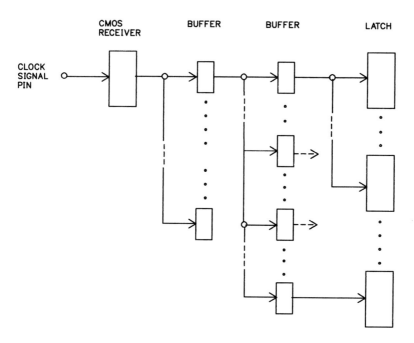

Figure 81. Standard on-chip clock distribution

The "logic chip dependent skew" originates in the receiving chips due to different clock powering schemes and process tolerances (see section 2.9.6). This type of clock skew adds a significant part to the total skew due to the performance difference between worst case and best case chips.

2.9.5 Logic Chip Clock Distribution

The clock pulses are used to control latches and arrays on the logic chips. The total capacitive load, which has to be driven by the clock is determined by the number of latches, their input capacitance plus the wiring capacitance.

There are two different clock distribution schemes used on the logic chips. They have been implemented to reduce the total capacitive load to reasonable values (2 pF per driving circuit).

2.9.5.1 Standard On-Chip Clock Distribution

The standard clock distribution (Figure 81) has a CMOS receiver and several buffers in sequence. This type of clock powering represents a capacitance of 5-10 pF at the module pin and may be driven by only one clock driver. It has the disadvantage of a relatively high path delay from the I/O pin to the ultimate target (latch input pin) caused by several buffer stages. Almost identical clock powering paths have a significant clock skew at the latch itself due to process tolerances (tracking) and differences in the capacitive load of the buffer stages.

The clock skew between two latches on different chips may increase to 6 ns, which is the difference in path delays between a worst case and best case chip. This fact leads to an earlier setting of the C1-clock vs. the B1-clock to avoid a clock overlap at the latches.

2.9.5.2 High Performance On-Chip Clock Distribution

This kind of clock distribution (see Figure 82 is especially designed for the performance critical paths of the processor. The logic contains only one amplifier stage (CLDR) between the chip I/O pin and latch and is designed with many parallel high performance inverters, which are dotted at the output. The "clock driver dot" is routed through the entire chip like a ring wire. In this case the I/O cell (clock receiver) contains an ESD device only. The signal is directly wired to the internal metal layer, avoiding any on-chip clock skew at the latches for that specific clock. Even the skew between two logic chips is almost negligible. This type of clock powering represents a high capacitive load at the module pin depending on the number of parallel inverters used for the clock distribution (between 15pF and 90pF). The load is driven by up to five clock chip drivers.

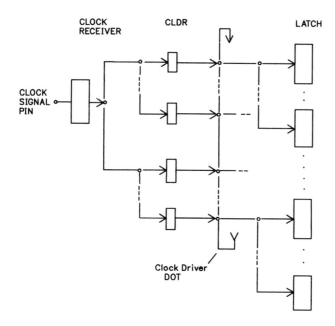

Figure 82. High performance on-chip clock distribution

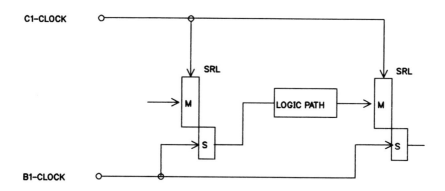

Figure 83. Basic clock triggering

2.9.6 Evaluation of the Clocking Scheme

The clocking scheme was developed based on the previous mentioned specifications. As shown in Figure 83, a logic path is triggered by a B1-clock controlled slave latch, and the logic outcome of the path is gated into a C1-clock controlled master latch "M". The logic path may contain chip crossings.

The positioning of the clock signals reflects the logic path delay, the minimum clock pulse width (MPW), the minimum gap between the C1- and B1-clock, the data hold time, and the appropriate clock skews.

The following values were used to develop the 80 ns cycle clock.

MPW = 7 ns, clock overlap = 0 ns, data hold time = 0 ns, logic path delay = 62 ns.

The clock skew in reference to the nominal clock position at the receiving module pin of the logic chip is defined with

+/- 3 ns for the clock pulse width
+/- 4 ns for the clock pulse distance

Figure 84 shows for the 80 ns machine cycle the nominal C1- and B1-clock setting together with the respective clock skews in reference to the clock pulse transitions.

Figure 85 shows the timing relationship of all clock pulses. This includes the C1 and B1 clock setting shown in Figure 84.

2.9.7 Clock Variation

An extensive timing verification was performed on the complete CPU card. For this the clock chip features were utilized to stress the timing of both the individual chips and the interchip timing. This was done to increase the confidence in calculated card timings, find marginal timings, correlate measurements with the delay calculator results (see "3.5 Timing Analysis and Verification" on page 177), stress clock skew specification, stress both critical chip internal paths, and critical chip to chip paths.

The measurements were done by using narrow or overlapping clock pulses until the chip set failed. Toward this end the clock chip was re-programmed to move the clock pulses into marginal positions.

As a result, chip set malfunctions occurred within expected margins. No unknown critical paths were detected. The correlation of measurements and calculations done for the CPU chip showed a difference of 2 ns (27 ns vs. 29 ns). The delay calculator tool proved to be sufficient for timing evaluations. Also the clock variation method is independent from cycle time. Therefore, shorter cycle times will show the same results.

These tests verified that chips within specified process tolerances will work as expected.

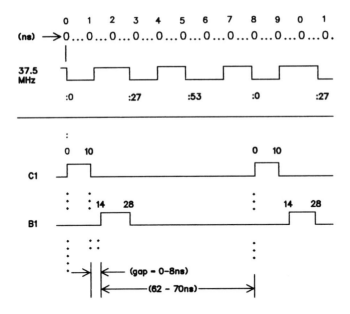

Figure 84. C1 and B1 clock setting

Clocking

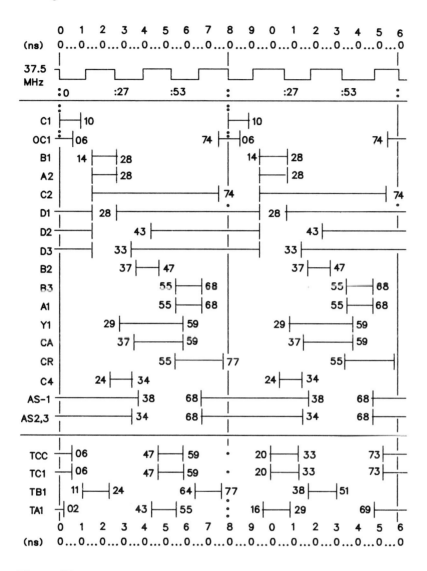

Figure 85. Clocking scheme for 80 ns cycle time

2.10 Processor Bus

Horst Fuhrmann, Hermann Schulze-Schoelling

2.10.1 Processor Bus Connections

The bidirectional Processor Bus interconnects the CPU, MMU, BBA and optionally, the FPU chips. A unit attached to the bi-directional Processor Bus is called a Bus-Unit (BU).

As shown in Figure 86 the Processor Bus consists of a 4 Byte wide Data Bus, a 5 bit wide Key-Status-Bus and some control lines, like Command-Valid, Bus-Busy, MMU-Busy etc. It operates in a handshake mode under control of an arbiter function. All bus operations are synchronously clocked.

The additional Support Bus (not shown) is described in "2.11.6 Support Interface" on page 141.

2.10.2 Processor Bus Implementation

Specific logic on the chips attached to the Processor Bus serves the bus operations. The bus transfer cycle is triggered by a B1-clock controlled slave latch. The receiving bus unit picks the data with a C1-clock controlled master latch.

The bus operations may be performed by two different chips within consecutive cycles, e.g. the CPU sends a fetch command to the MMU in the first cycle and the MMU responds with the data in the following cycle. Since the change of the bus ownership from one cycle to the other may cause an orthogonal driver switching problem, all drivers of the bi-directional bus signal lines are forced into the high impedance state for a certain time by a driver-enable-clock (D1-clock). The activation of the dedicated bus unit drivers is controlled by a "bus-enable-latch" together with the ongoing D1-clock. This is shown in Figure 87.

The Capitol Chip Set is designed for two bus masters: CPU and BBA. The arbitration function is implemented on these two bus masters by assigning the lowest priority to the CPU chip. It has a default bus request and therefore does not provide a bus request pin. The signal BUS_GRANT CPU is the inverted signal BUS_REQ BBA. The BBA has the highest priority. Its BUS_REQ BBA signal is inverted and wired as BUS_GRANT CPU to the CPU.

Switching the Processor Bus ownership from a high priority bus master to a lower priority bus master does not cause a Processor Bus cycle loss. There is no bus unit (chip attached to the Processor Bus) interleaving on the Processor Bus. The Processor Bus is not granted to another bus unit as long as the data transfer (fetch/store) of the current command is not completed.

Processor Bus

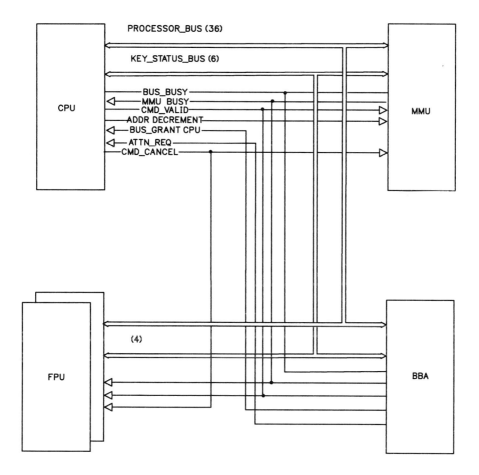

Figure 86. Processor bus

Commands are executed sequentially. This means that the current command execution must be completed before a new command can be issued. As an example, after a bus unit has issued a "store command" it cannot issue another MMU command or message command as long as its store command is not yet completed.

2.10.3 Processor Bus Operation Example

Figure 88 shows a 5 cycle CPU request to the MMU, e.g. a "Fetch 12 Byte" operation, followed by a one cycle BBA control to a Bus Unit and the beginning of another CPU conversation. The letters C, M, B indicate that the CPU, MMU, or BBA chip puts a signal on the corresponding control line. The functions of some of the control lines are:

CMD_VALID is a bi-directional control line. The bus master uses the

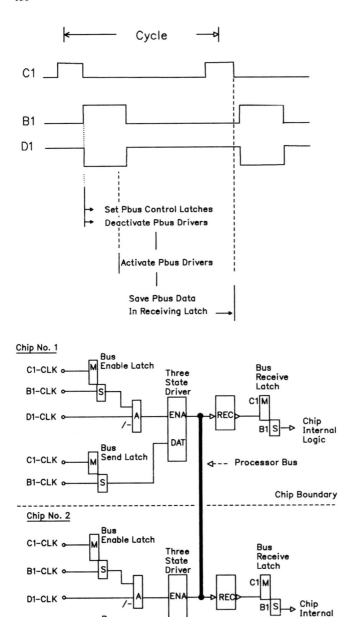

Figure 87. Principle of the processor bus driver logic

Processor Bus by activating CMD_VALID and setting data on the Processor Bus provided its BUS_GRANT is active and BUS_BUSY was inactive during the preceding cycle. Each Bus Unit is triggered by CMD_VALID to receive and decode the command placed on the bus.

BUS_BUSY is a bi-directional control line that is activated by a bus master, if it uses the Processor Bus for longer than one cycle. This signal indicates that the Processor Bus will be used in the next cycle. The MMU or BBA will activate BUS_BUSY if it cannot send back the data right after the command cycle. It gets inactive one cycle before the bus usage will be terminated.

MMU_BUSY is a uni-directional control line driven from the MMU or the BBA to all bus masters. For a fetch operation, an inactive MMU_BUSY after the command cycle indicates valid data on the Processor Bus.

During the machine cycles in Figure 88 the following events occur:

2. The CPU gets BUS_GRANT-CPU for one cycle.

3. CPU starts a conversation by activating CMD_VALID (see "2.2.6.2 Microinstruction Types and Formats" on page 34). In parallel, the command is put on the Processor Bus, and the storage key on the KEY_STATUS_BUS. Because the Processor Bus will be utilized for more than one cycle, the CPU activates BUS_BUSY for one cycle.

    In the same cycle BBA requests the next (free) Processor Bus cycle.

4. BUS_BUSY in this and the following cycles indicates that the MMU has decoded the command length count and it activates BUS_BUSY for three more cycles. As BUS_BUSY determines the bus usage one cycle ahead, the bus is used up to and including cycle 7 (one wait cycle and 3 data transfer cycles). MMU-BUSY active indicates that no valid data are on the bus.

5. MMU_BUSY inactive indicates that valid data are on the bus.

7. This is the last data transfer cycle for the MMU, where it also puts the status on the KEY_STATUS_BUS, and the arbitration cycle for the next bus event. BBA must control the Processor Bus in the next cycle: BUS_BUSY is inactive, BUS_REQ_BBA active.

8. BBA deactivates BUS_REQ BBA and BUS_GRANT CPU is activated by the CPU. BBA controls the Processor Bus. As BUS_BUSY is not active, this is the arbitration cycle for the next bus cycle.

9. CPU takes control of the Processor Bus for one ore more cycles.

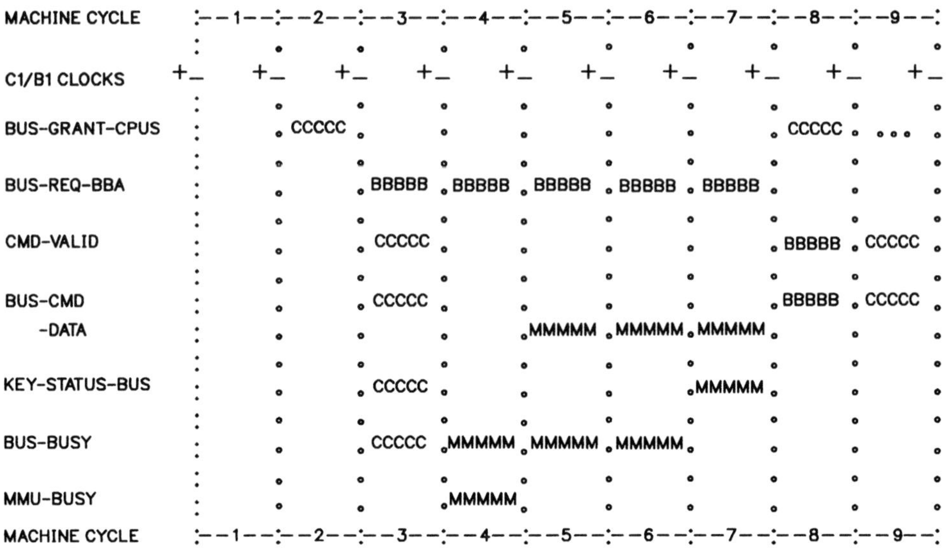

```
MACHINE CYCLE    :--1--:--2--:--3--:--4--:--5--:--6--:--7--:--8--:--9--:
                 :    .     .     .     .     .     .     .     .     .
C1/B1 CLOCKS     +_   +_    +_    +_    +_    +_    +_    +_    +_    +_
                 :    .     .     .     .     .     .     .     .     .
BUS-GRANT-CPUS   :    . CCCCC .    .     .     .     .  . CCCCC .  ... .
                 :    .     .     .     .     .     .     .     .     .
BUS-REQ-BBA      :    .     .BBBBB .BBBBB .BBBBB .BBBBB .BBBBB .    .     .
                 :    .     .     .     .     .     .     .     .     .
CMD-VALID        :    .     .CCCCC .    .     .     .     . BBBBB .CCCCC .
                 :    .     .     .     .     .     .     .     .     .
BUS-CMD          :    .     .CCCCC .    .     .     .     . BBBBB .CCCCC .
   -DATA         :    .     .     .     .MMMMM .MMMMM .MMMMM .    .     .
                 :    .     .     .     .     .     .     .     .     .
KEY-STATUS-BUS   :    . CCCCC .    .     .     .MMMMM .    .     .     .
                 :    .     .     .     .     .     .     .     .     .
BUS-BUSY         :    . CCCCC .MMMMM .MMMMM .MMMMM .    .     .     .     .
                 :    .     .     .     .     .     .     .     .     .
MMU-BUSY         :    .     .     .MMMMM .    .     .     .     .     .
MACHINE CYCLE    :--1--:--2--:--3--:--4--:--5--:--6--:--7--:--8--:--9--:
```

Figure 88. Processor bus timing

2.11 Reliability, Availability, Serviceability

Peter Rudolph

2.11.1 Overview

The terms Reliability, Availability, and Serviceability, short hand RAS, are widely used acronyms within the community to characterise the features that make a part (hardware or software) of a data processing system less failure prone as both the end user and the service technician (customer engineer) sees it. The catalogue of characteristics employed to assure good RAS performance encompasses many individual items. The RAS "strategy" of a new product is the agreed upon set of hardware and software characteristics that assure superior RAS performance.

One of several RAS measures is the number of repair actions (RA) for a given machine type during a stated time period (usually a month). This refers to the frequency, at which a customer engineer is being called to service (repair) a machine. For a population of a particular machine type, RA's follow a learning curve as indicated in Figure 89, characterized by the 3 points A, B, and C. These values are specified at the start of a new project, reported back to the development laboratory by the field engineering organisation, and evaluated at time C. Repeatedly during the development process, at predetermined checkpoints, the RAS facilities are reviewed as to their adequacy to meet the objec-

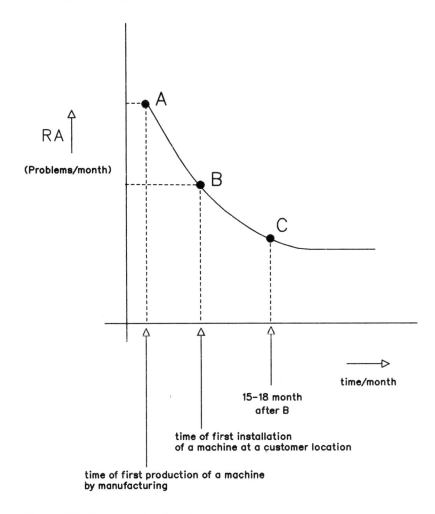

Figure 89. Repair action learning curve

tives. If at time C this is not the case, causes for missing the objective are analysed, and, as a rule, corrective action is taken.

An important factor in developing the RAS strategy is the "IBM Corporate Instruction 105". It states, as indicated in Figure 90 "The quality of a new product "2" at the time of first installation must be superior to that of the currently available product "1", which it replaces. This implies a continuous progress in RAS characteristics from one product generation to the next, and frequently unusual measures to achieve the goals thus established. Because of this, IBM products historically have been above average in regard to RAS, and the Capitol chip set is no exception. The RAS features are significantly more comprehensive than those found in other available 32 bit microprocessors. For example, 25-30 % of the silicon real estate is devoted to RAS and testing functions.

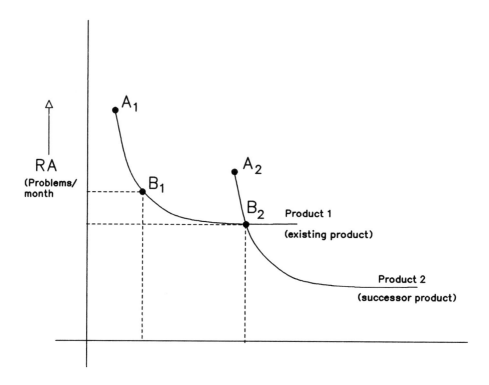

Figure 90. Product quality goal setting

2.11.2 RAS Strategy and Requirements

In the past, products suffered in quality when functions with high RAS dependencies were implemented mainly under operation considerations. To achieve optimum RAS characteristics, corresponding facilities have to be a part of the original design.

An example is the Initial Micro-program Load (IML) function. It loads the control program code into the Control Store (CS) and non-S/370 memory. RAS aspects verify the loading, and analyse occurring problems. The RAS part exceeds the operational part and thus dominated the implementation of the IML function.

The Capitol chip set is designed for use with or without the support of an auxiliary processor, the so called Service processor (SP). Especially, functions otherwise executed in the Service Processor that cannot be omitted in the Capitol chip set environment, are implemented and executed in the available logic. An example is the Data Local Store initialization after power on, error handling after any check, hardware testing, IML, etc.

The Capitol chip set is divided into 3 domains with different error reactions:

- Errors in the processor nucleus stop the clock.

- Errors in the memory and bus adapter area initiate a trap to micro code (any check trap).

- Errors within the I/O transfer path (control and data) cause an exception interrupt at the end of a current S/370 instruction.

This arrangement allows the usage of the processor intelligence to handle peripheral error situations by microprograms.

Error situations in the processor nucleus are cleared and retried under hardware control. A "wait" period of approximately 2 seconds after clock stop by a checker allows physical intermittent error causes to settle before the processor nucleus is reset for an error free restart of microcode execution (check restart).

A second error, while still in an error handling phase, stops the clock again and blocks the restart after check. The "check stop" line is activated signalling a solid error situation within the Capitol chip set nucleus. This line may directly turn on a red light indicator on a panel. The Capitol chip set leaves the check stop state by "power on reset", re-IML, and system or "load reset".

The checker information and other machine error data and states are retained and readable by micro instructions after check restart. This error status information is kept in latches that are reset under hardware control during power-on, IML, and if a S/370 system reset or load reset (IPL) is requested. The Clock chip provides hardware inputs for all the different reset signals so that they may be directly connected to panel push buttons.

2.11.3 Initial Chip Set Start and Loading

To meet IBM VLSI chip production and test requirements, all chips of the Capitol chip set are designed according to the Level Sensitive Scan Design rules. For details see "5.1.5 LSSD (Level Sensitive Scan Design)" on page 248.

The basic idea of LSSD is to split functional sequential digital networks for testing into pieces of combinatorial networks. To cut the feedback loops in sequential networks, all storing elements (latches) in the logic flow are implemented as master/slave latches that also work as shift register latches (SRLs) connected into one or more chains (SRL chains, long shift registers). Figure 91 shows some details. The serial inputs and outputs, and the controlling clock signals are accessible at chip pads or pins. Test patterns for the combinatorial logic part can now be scanned in from an external source, and the resulting pattern can be scanned out after forcing a process step. The SRLs can be considered as pseudo input or output pins.

Imbedded arrays are normally assembled from storage elements. They do not provide a serial path. Due to their regular structure they are tested by control and data patterns serially set up in SRLs directly surrounding the arrays.

All chip set functions depend upon timing (clock) pulses supplied by the Clock chip. The three general clock pulse types (clocks) which control the SRL based logic are:

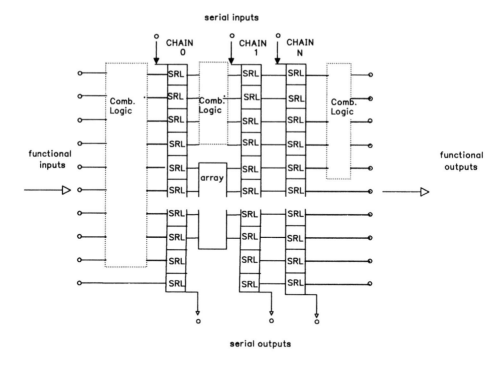

Figure 91. SRL chain structure

- Process Master (C) clock,
- Shift Master (A) clock,
- Slave (B) clock, common for processing and shifting.

Chip set processing stops (clock stop, stop hardware functions) by suspending the supply of certain C clocks, called "stoppable" clocks. Continuously running C clocks control functions, which have to proceed during a functional stop state like main memory refresh, time of day counter update, etc. B clocks, controlling the SRL internal transfer from master to slave latches, never stop.

The shifting of SRLs (for serial reset or support access) requires stopping of the associated C clocks.

When power has stabilized after power on, the "power-on reset" signal triggers the initialization of the clock chip. The clock generation latches on the clock chip are set by internal or external patterns as defined from two selection inputs. The run control is forced into its initial state ready to control the power on reset sequence for the other chips.

The "chip set de-gate" line is activated and is used to fence external chip set interfaces as long as the chip set is in an uncontrolled state. To activate the signal as early as possible (when the raising voltage reaches switching level), it connects to ground potential via a resistor. The line is switched off by a CPU microinstruction, when chip set functions are guaranteed after initialization and loading. The chip set de-gate is also activated, if the chip set enters an undefined state due to check stop or clock check.

As soon as the clock generation logic is initialized, the shift clocks (A clocks) and the slave clocks (B clocks) start. During this phase, all C clocks are still blocked. Simultaneously, the reset control lines RESET_COND and UIF_RESET are activated (see Figure 92) to gate the power on reset shifting in the logic chips for 2048 shift cycles.

When the serial reset is completed, the clock generation starts to supply all continuously running C clocks. The stoppable clocks still remain shut off (clock stop state). The subsequent wait period of another 2048 cycles ensures stabilization of the main memory refresh function, which runs under control of continuous C clocks.

After the stabilization wait phase, the stoppable C clocks are provided in a defined clock start sequence. The chip set begins execution of micro operations (chip set running). The very first operation fetches 64 bytes from a non-volatile microcode source EPROM into the microinstruction buffer area in the control store. The control conditions to execute this initial "fetch 64 bytes" from EPROM were forced by the serial reset before. All checkers are active (hot) to secure this initial function.

The microinstruction address is forced to x'0000' to start instruction execution at the first instruction in the microinstruction buffer (in CS). This instruction address is expanded by a forced base address of all zeros to form a three byte address (16 MB range) of all zero, sent out to select the first 64 Byte page from the EPROM (start block of microcode), which is transferred into the buffer and then executed.

The "Fetch 64 Byte" is a general function, which either accesses the main memory (normal) or the EPROM. To differentiate between both access, the EPROM "mode latch" is set. The latch status controls the (low order address bit on Processor Bus) MMU and BBA to select the microcode source.

As long as the EPROM mode latch is set, microcode is only executed out of the microinstruction buffer in the control store. When branching or sequential address increment results in an address pointing outside the 64 bytes just in the buffer, the respective microcode page is loaded prior to further execution. The EPROM mode latch is reset by a microinstruction to terminate execution of EPROM code.

The EPROM code or data provide correct byte parity. Besides the "Fetch 64 Byte" from the EPROM, data can be read into a Data Local Store (DLS) register, four bytes at a time. The EPROM may contain microcode, data, control information, S/370 programs, etc., which are finally stored into the main memory or control storage.

A two Byte "Write Control Store" microinstruction transports Data Local Store (DLS) information into a selected Control Store (CS) position. This instruc-

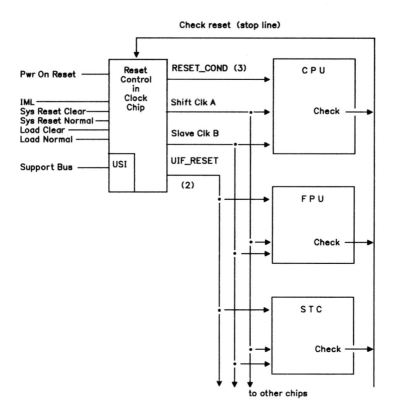

Figure 92. Chip set reset lines

tion is used to load microcode from the EPROM into the CS. The "Read CS" microinstruction places two byte of CS data into a DLS register. Read CS operations are used to read back CS content to verify loading. The Write/Read CS operations are not restricted to loading, but are used whenever information has to be saved and retrieved within the processor nucleus.

The Capitol chip set resets are controlled from the central run control on the clock chip. Five reset control lines (UIF_RESET, RESET_COND, see Figure 92) pass the reset type and the reset source to the logic chips. The non-zero state of these line groups are decoded and gate the shifting of a forced binary "0" into serial paths provided by the SRL implementation described in "2.11.6 Support Interface" on page 141 (Serial Reset). Within the SRL chains, the latches are sequentially arranged according to their reset type. The different resets are achieved by controlling how far the forced "0" propagates into an SRL chain. The reset control is generally active for 2048 shift cycles designating the possible length of a reset chain.

2.11.4 Error Detection

Checking circuit (Checker) implementations make the consequences of failures in the covered logic predictable. Checkers allow to develop and implement controlled error recovery and repair actions. Past experience indicates that even single failures occurring in unchecked logic are frequently treated and reported to the development group as design problems and cause (erroneously) corrective development activities. These failures cannot be distinguished from design failures. The implementation of checkers is part of the responsibility of the individual logical designers.

Error detection is guaranteed by a high level of check coverage. Check coverage is defined as the fraction of circuits that when failing cause at least one checker to respond. The Capitol chip set check coverage is better than 95%. The coverage is counted using CMOS devices or equivalent elements as a base, and is expressed as an average value for each chip. To prevent over-emphasis, check coverage of arrays is considered and counted as if the array would consist of one position only (e.g. one byte).

In the future a tool is required that directly assigns and counts the number of elements belonging to the error domain of a checker. Currently this must be done manually by the hardware designers.

Hardware checkers cannot be disabled, when the Capitol chip set is working at the functional level (micro code level), the correct operation of which they ensure. The "hot checker" strategy allows a verified execution of micro coded maintenance and support functions within the endangered processor area, when a chip set problem occurred previously. For bring-up and debugging activities, the effect of active hardware checker can be suppressed (check override, check disable) for a single checker or a checker group via the Unit Support Interfaces (see "2.11.6 Support Interface" on page 141) on the chips. However, the check latches associated with the checkers are set, and subsequent check reactions are inhibited.

An assortment of different checkers and methods provide check coverage for the various Capitol chip set areas:

- The chip set data, address, and instruction flow employs byte parity checking. The parity is maintained through all data and control storages or buffers (CS, cache, DLS, etc.). Parity prediction and correction secure operations of the CPU ALU and the various incrementing and decrementing devices.

- Main memory RAS functions on the STC chip include check bits (Hamming code, ECC) for single error correction and double error detection (SEC / DED), retry, scrubbing, redundancy bit control, and others. "2.5.3 Data Integrity" on page 85 discusses these functions in detail.

- The FPU chip is duplicated. The second (shadow approach) chip compares all outputs of the first one. The FPU chips check and generate the processor bus parity.

- Various special decode and plausibility checkers expand the error detection on the control logic. The CPU time out checker serves as a global control checker. It monitors the alterations of the control store address register. A lack of activities over a certain period of time activates the checker. It stops the clock, and the subsequent reset and restart clear a possible stuck or hang situation. The restart gives control to adequate service routines.

- The clock chip is secured by a clock checker recognizing violation of clock pulse generation. It stops the clock supply immediately and forces a re-initialization of the clock pulse generators as after power on. This re-initialization causes power on reset similar to the normal power on followed by the processor restart. The fact of the clock check occurrence is signalled via the clock check back-up line to the CPU chip where it can be read and reset by microinstruction.

- Inbound data on the Processor Bus and STC Bus are byte parity checked. For outbound data on these busses byte parity is generated. The arrays for the Translation-Lookaside Buffer (TLB), Cache Directory, Cache, and Key Store are byte parity checked. Checking is done to avoid double TLB selection and multiple late selection of Cache columns. The ACB register (see "2.4.3 MMU Chip Data Flow" on page 56) is Byte parity checked.

- Correct operation of all checkers is verified during each power on sequence by activating special controls (Checking the Checkers).

2.11.5 Machine Check Handling

Machine check interrupt is a S/370 architected function. It interrupts S/370 processing and signals that error information is available in the S/370 page zero area (machine check logout area), containing information like the machine check interruption code, failing storage address, register save area, extended interruption information, and machine specific logout areas. The operating system saves the information and performs error relevant recovery actions. The machine check interruption code is a collection of information bits on error type and severity expanded by the validity states of vital S/370 oriented system data and states. This code is generated by the error handling microprogram that gains control after initial check restart (clock stop, serial reset), if "Any Check" trap has interrupted the microinstruction sequence, or when a check attention exception (attention interrupt) is encountered between two S/370 instructions.

The error handling microprogram collects and investigates the error status saved in hardware registers (back-up registers, check registers) accessible to microinstructions.

Reliability, Availability, Serviceability   141

For validation, S/370 specific data in registers and main storage is accessed via checkers in the data path. A renewed check indicates that the information is damaged and invalid. The respective validation bit in the S/370 machine check interruption code is set accordingly.

In catastrophic situations, the microprogram enforces the chip set into the check stop state (clock chip check stop line active) by a microinstruction ("Stop Immediate"). Any unexpected check activation (except during validation) while running the error handling microprogram results in "Check Stop". The "First Error" latch is reset by a microinstruction just before performing the S/370 machine check interrupt, which passes further operation control back to the S/370 interpretation.

The ability of an operating system to use this information, and to recover from random, intermittent errors depends on implemented functions. For three widely used S/370 operating systems the values achieved under a given set of circumstances were in 1985:

```
VSE(Release 1) = 60%
VM             = 70%
MVS            = 86%
```

These figures have increased in more recent releases.

2.11.6 Support Interface

2.11.6.1 Unit Support Interface Description

For hardware and micro code support in the field, manufacturing, and for initial bring-up and functional testing, a Unit Support Interface (USI) is implemented as an integral part of the CPU, MMU, FPU, BBA and Clock chips. It connects to the serial, five-line Support Bus (SB) of two bit lines (in/out) and three control/ clock lines, (see Figure 93).

The data in/data out lines shift address or data bits from/to the unit support interfaces. The address line indicates whether address of data information is shifted. The set pulse line is activated to indicate that an information group (e.g. an address or a data byte) has been completely transferred. The shift gate line synchronizes shift clock pulses.

An actual system implementation with the Capitol chip set may or may not utilise the Support Bus and the USI interface. In the ES/9370 system the Support Bus is driven by a Support Bus Adapter (SBA) that is part of a standard IBM Personal Computer. The PC functions as a service processor, and allows PC programs to manipulate hardware details in the attached chips.

The PC and the Capitol chip set CPU card are interconnected using the standard PC printer cable. The cable has a Centronics plug on the printer side which connects to the processor card. Four receiver and one driver circuits on

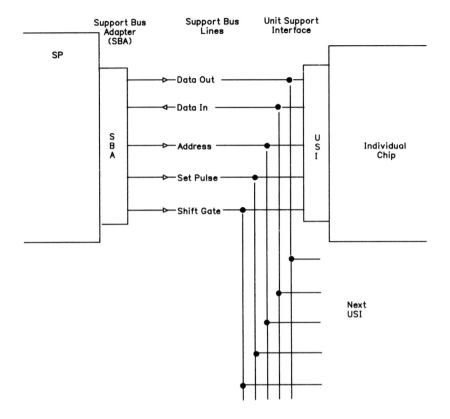

Figure 93. Support bus interface

the Processor card receive and distribute the Support Bus to the USIs on the individual chips, see Figure 94. The receivers and the driver secure the Capitol chip set modules against hostile interface conditions.

2.11.6.2 Unit Support Interface Operation

The USI is a logic function on each of the chips. It serves as a backdoor interface to the Capitol chip set modules. The interface operates independently of the status of the logic so that support is available in unpredictable error or design problem situations.

The main function of an USI is to select shift latch (SRL) chains, which are integrated in the chip logic to meet the LSSD design rules, and to interconnect them to the Support Bus.

To read and/or alter the information locked in the SRLs, the bits in a selected SRL chain are shifted via the "bit in" and "bit out" lines, through a "window" register on the SBA side, and back into the selected chain (ring shift). By controlling the amount of shift pulses, information from any part of a chain can be

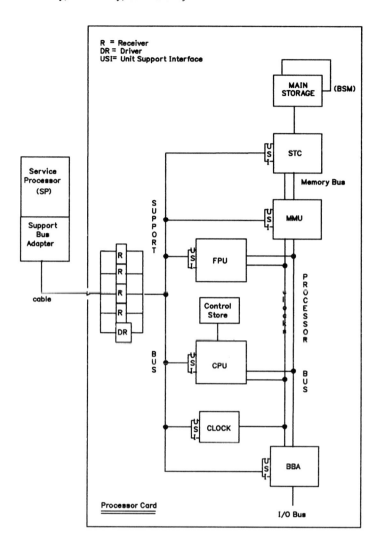

Figure 94. Support bus interconnection

positioned in the "window" so that it can be read or altered in parallel. A "360 Shift" is that shift count required to return all bits into their origin positions.

Any bit in a chain can be specified by the number of shift pulses required to shift it out of its chain. To avoid unnecessary adjustments and sorting activities of chain data, functional bit groups are kept together and not sprinkled all over. E.g., the following is the convention for the sequence of bits in a data byte.

-|7|6|5|4|3|2|1|0|P|- (out)

A shift out starts with the parity bit (P, if any) followed by bit position 0, etc.

Part 3. Logic Design Tools

3.1 Logic Design System Overview

There are 2 major steps in generating the data, from which the masks for a VLSI chip can be generated, see Figure 95.

The first step, the "logic design", consists in generating the interconnection patterns for the logic circuits being used, e.g. NANDs, NORs, latches, macros, etc.. Within IBM the result is a design in a proprietary gate level description language, called BDL/S (Basic Design Language for Structure). For a description, see [MAIS]. Part 3 describes the "Logic Design System", i.e. the tools that are used to generate the BDL/S design.

The second step, the physical design, uses BDL/S data to generate a RIT (Release Interface Tape) file, that is used to directly drive the mask making equipment. The tools used in this step are described in Part 6.

The Capitol chip set was designed with the help of Silicon Compilation. The logic on the CPU and MMU chip was exclusively designed using the higher level "Design Language" described in "3.2 Hardware Design Language" on page 148, and logic synthesis software described in "3.3 Logic Synthesis" on page 158. A simple Design Language example is shown in Figure 96

In principle, as shown in Figure 97 there are 2 alternative paths to generate gate level data: the traditional graphic approach of entering logical building blocks including their interconnections, and the newer method of silicon compilation. Analogous to the coexistence of higher level language and assembler code in large software engineering projects, both approaches may have a future in VLSI designs.

The Design Language source level description of the logic is compiled with the help of the Design Language Compiler. The compiler output is piped (serves as the input) to the Logic Synthesis software that generates BDL/S. This step is shown on the right side of Figure 98. In addition, the same Design Language description is used for logic design verification via logic simulation. For this the design language source, together with a description of its input/output environment described in a "Behavioural·Language" is compiled by the same Design Language Compiler (using a different compile option) into S/370 machine code. This, together with separately generated test cases is used by the simulator software to verify that the Design Language description of the logic design meets its functional objectives (see left side of Figure 98 and the description in "3.7 Logic Simulation" on page 190).

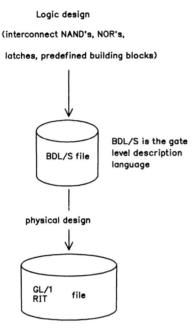

Figure 95. Major logic design steps

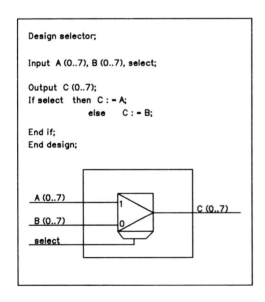

Figure 96. Design language example

Logic Design System Overview

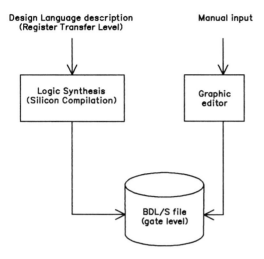

Figure 97. Two alternative ways to generate BDL/S data

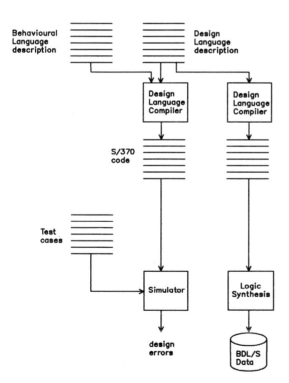

Figure 98. Logic design and verification overview

3.2 Hardware Design Language

Wolfgang Rösner

3.2.1 Overview

The design of the Capitol chip set used a new hardware design language, in the following referred to as Design Language. It is a structural language that was designed to serve as an input medium for both logic simulation and synthesis. The corresponding compiler generates simulation code for the Boeblingen Mixed Level Design Verification Simulator (see "3.7 Logic Simulation" on page 190) and input for IBM's Logic Synthesis System (LSS) [DAR2].

A designer uses the Design Language to describe hardware hierarchically in 3 layers or levels from the system level down to the macro level (logic gate level) (Figure 99). The Design Level is equivalent to the Register-Transfer-Level. The System Level is used to specify the linkage of hardware blocks which are described on the Design Level. This allows the assembly of Chip-, Card-, Board- or even total System descriptions.

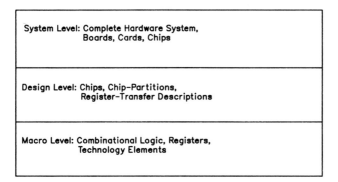

Figure 99. Design language description layers

On the Design Level the logic design is entered as a combination of I/O net declarations, Boolean equations, and higher level constructs such as IF, CASE and PLA-like statements.

A powerful macro facility is used, among others, to include technology dependent parts. These elements which are just repertoire blocks or black boxes for LSS are associated with a simulation description on the Macro Level.

The Design Level of the language allows the specification of logic functions either by macro invocation and linkage, or by the use of Boolean equations, or other technology-free statements. This allows specifications of designs either on a technology independent level, on the technology circuit level, and a continuum of all mixes between these two extremes.

Thus the designer guides the logic synthesis process for all parts of his design. Technology independent parts are synthesized using the full set of LSS's Boolean and technology dependent transformations. Technology parts described by macros are left untouched. This means that for logic synthesis the Design Language provides all means for a pragmatic design methodology which allows rapid and implementation independent design of non-critical parts and hand-layout design of the critical parts of a VLSI chip.

The Design Language provides attributes for signals. This allows the compiler to apply certain design rules checks very early in the design cycle.

There are features in the language which are directed to logic synthesis alone, e.g. physical attributes or requested arrival times. Others are useful for the simulator only, e.g. Assertions or the simulation descriptions of the technology blocks.

3.2.2 The Design Level

Designs are pieces of hardware described on the Register-Transfer-Level or Gate-Level which normally represent chip partitions or complete chips.

Reasons for chip partitioning into designs are project management and data management requirements (esp. for VLSI projects) and LSS run-time and capacity limits (see "3.3 Logic Synthesis" on page 158).

A design represents a piece of hardware with

- Inputs and outputs (which may be the chip pins)
- Combinatorial logic
- Hardware elements which cannot be described as combinatorial logic.

The corresponding Design Language language elements are

- Input/output signal declarations
- Statements or PLA-like tables
- Macro references

Figure 100 shows an example for a design. It is a simple ALU description using a special adder macro.

Input/Output signal declarations (connectors) are mandatory to describe the interface of the design. I/O connectors and nets used locally in a design may be single bit or bundled signals. Local nets need not be declared. It is obvious from the example that the Design Language is a **descriptive network specification language**. The sequence of the statements is immaterial.

Figure 100 shows three statements representing their statement types. The adder which is central to the ALU is specified by invocation of a macro. It may be a technology element or a user macro. Note that the syntax of the lan-

```
DESIGN ALU

  INPUT   A_REG(0..7), B_REG(0..7),
          COMMAND(2..4);

  OUTPUT  S_OUT(0..7), CARRY_OUT;

  DEVICE  MY_MINI_ALU: ADDER (WIDTH = 8)

  ALU_IN1(0..7)....| IN1
  ALU_IN2(0..7)....| IN2
  CARRY............| C_IN
                   | SUM    |S_INT(0..7).......
                   | C_SUM  |CARRY_OUT.........;


  CASE COMMAND(3..4) OF
                                    /* ADD */
      B'00': ALU_IN1(0..7) := A_REG(0..7);
             ALU_IN2(0..7) := B_REG(0..7);
             CARRY         := 0;
                                    /* SUBTRACT */
      B'01': ALU_IN1(0..7) := A_REG(0..7);
             ALU_IN2(0..7) := ¬ B_REG(0..7);
             CARRY         := 1;
                                    /* INCREMENT */
      B'10': ALU_IN1(0..7) := A_REG(0..7);
             ALU_IN2(0..7) := 1;
                                    /* DECREMENT */
      B'11': ALU_IN1(0..7) := A_REG(0..7);
             ALU_IN2(0..7) := X'FF';
             CARRY         := 0;

  END CASE;

  S_OUT(0..7) :=
    (COMMAND(2) GATE S_INT(0..7))
  | (¬COMMAND(2) GATE (A_REG(0..7) & B_REG(0..7)));

END DESIGN
```

Figure 100. Simple ALU design

guage allows the graphic-like appearance of macro block invocations in the design source.

The S_OUT bundle is driven by combinational logic and is specified by an assignment statement. Two logic values are possible assignment values. For bus resolution functions (i.e. "dotfunction") there are three strength levels provided. The case statement whose output controls the data input and operation of the adder represents a selector.

It should be noted that all the logic elements used in the ALU description are independent of any technology. The LSS processor will do the optimized mapping to the technology level.

This is different for the adder macro which is just invoked and provided with input/output signal connections and the width parameter. Macros are expanded during the analysis and compilation of a design. For logic simulation a behavioral description must be the result of that process. For synthesis it is possible that the macro expansion results in a black box, i.e. a technology circuit specification which is then left untouched. Macros are discussed in more detail in "3.2.4 The Macro Level" on page 154.

3.2.3 Design Rules Checks

Signals appearing in the I/O signal declaration list can have attributes which carry additional logical or physical information about the signal. Logical attributes are used for design rules checks very early in the design cycle. These rules allow the checking of a subset of the LSSD design rules (see "5.1.5 LSSD (Level Sensitive Scan Design)" on page 248). Physical attributes are directed to the synthesis- or physical-design process.

The Design Language supports both LSSD and non-LSSD logic designs. LSSD designs should make use of attributes. Attributes are assigned to I/O signal declarations (or register declarations in a macro). Most important are the logical clock and latch attributes.

- MASTER_CLOCK/MASTER_LATCH
- SLAVE_CLOCK/SLAVE_LATCH
- SHIFT_CLOCK

Attributed signals are treated similar to typed variables in programming languages. Logical attributes are propagated through combinational logic networks, i.e. a signal having an attributed signal in its cone of influence inherits this attribute. Rules for attribute inheritance allow the compiler to check for design errors: (for single clock designs)

- Combinational logic driven by more than one clock signal
- Master-to-Master logic
- Slave-to-Slave logic
- Master_clock gated with master_latch output
- Slave_clock gated with slave_latch output

These rules are being extended for multiple clock designs. The clock attributes are also important for simulation code generation (see "3.7.3 Compilation

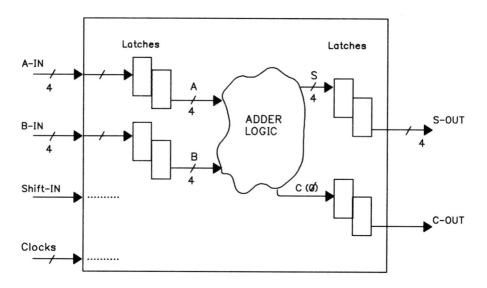

Figure 101. 4 bit adder enclosed in shift register latches (SRL's)

Techniques" on page 196) and for simulation run-time checks (e.g. master_clock and shift_clock signals must not be active simultaneously).

Figure 101 and Figure 102 show the specification of a 4 Bit adder logic enclosed in shift-registers-latches (SRL). The logical clock attributes in the ADDER input list and in the corresponding input list of the SRL macro guarantee that the proper clock signals are connected to the latches. The signals S(0..3) and C(0..3) are SLAVE_LATCH influenced signals because they are derived from A(0..3). There is no problem to feed them again into master latches whereas the compiler would report a design error if MASTER_OUT was the source of A(0..3).

An example for a **physical attribute** which is passed to LSS is shown in figure Figure 103.

The AT attribute specifies arrival and departure time. Since simulation is done in a zero delay mode a delay calculator is used for timing analysis (see section "3.5 Timing Analysis and Verification" on page 177). Thus timing information is physical information in the Design Language. It is passed to LSS to support timing path minimization (see section "3.3 Logic Synthesis" on page 158). For an input net, AT specifies the arrival time as an information. For an output net, it specifies the departure time as a requirement.

In design XX, signal A leaves the latch at slave clock time, i.e. at cycle time zero. The AT attribute tells the LSS processor, that the signal should arrive at the output of XX not later than 25 ns after that (i.e. the internal latch delay plus the combinational logic delay must not be greater than 25 ns). In design YY, the AT attribute tells LSS, that this signal is 25 ns later than the start of

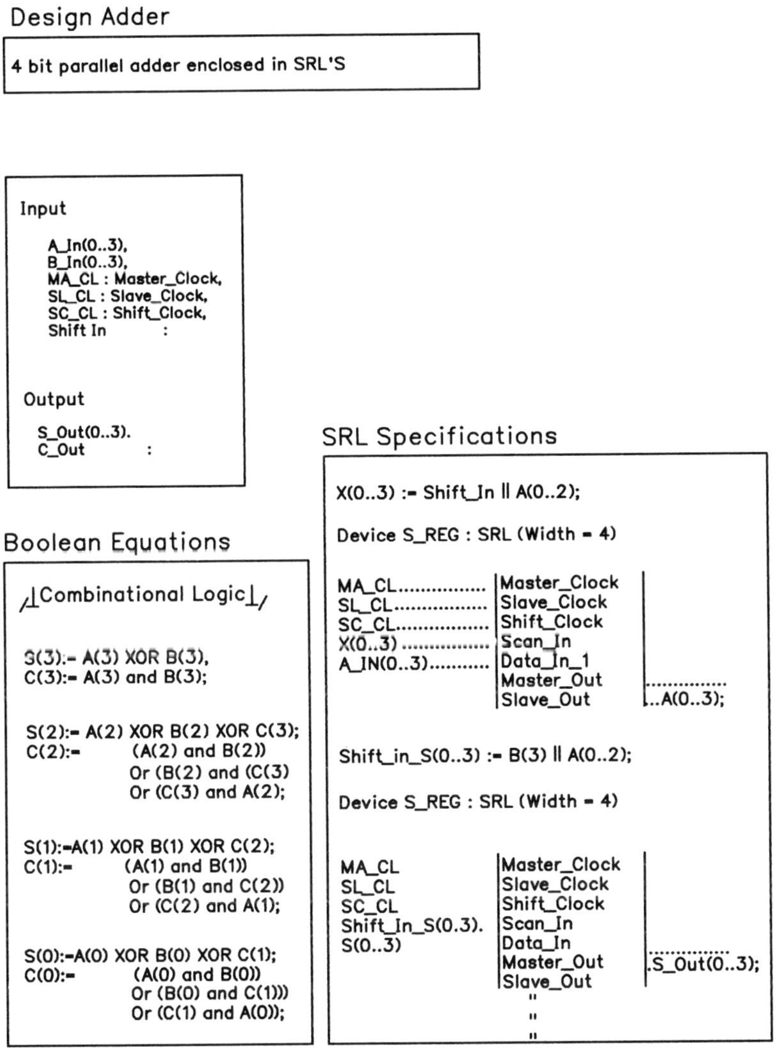

Hardware Design Language

Design Adder

| 4 bit parallel adder enclosed in SRL'S |

Input

A_In(0..3),
B_In(0..3),
MA_CL : Master_Clock,
SL_CL : Slave_Clock,
SC_CL : Shift_Clock,
Shift In   :

Output

S_Out(0..3).
C_Out   :

Boolean Equations

⊥Combinational Logic⊥

S(3):= A(3) XOR B(3),
C(3):= A(3) and B(3);

S(2):= A(2) XOR B(2) XOR C(3);
C(2):=   (A(2) and B(2))
         Or (B(2) and (C(3))
         Or (C(3) and A(2));

S(1):=A(1) XOR B(1) XOR C(2);
C(1):=   (A(1) and B(1))
         Or (B(1) and C(2))
         Or (C(2) and A(1);

S(0):=A(0) XOR B(0) XOR C(1);
C(0):=   (A(0) and B(0))
         Or (B(0) and C(1)))
         Or (C(1) and A(0));

SRL Specifications

X(0..3) := Shift_In || A(0..2);

Device S_REG : SRL (Width = 4)

MA_CL........	Master_Clock
SL_CL.........	Slave_Clock
SC_CL.........	Shift_Clock
X(0..3).........	Scan_In
A_IN(0..3).....	Data_In_1
	Master_Out
	Slave_Out ..A(0..3);

Shift_in_S(0..3) := B(3) || A(0..2);

Device S_REG : SRL (Width = 4)

MA_CL	Master_Clock
SL_CL	Slave_Clock
SC_CL	Shift_Clock
Shift_In_S(0..3).	Scan_In
S(0..3)	Data_In
	Master_Out .S_Out(0..3);
	Slave_Out
	"
	"
	"

Figure 102. 4-bit adder design language example

the cycle. LSS will subtract those 25 ns from the time available for the combinational logic in YY.

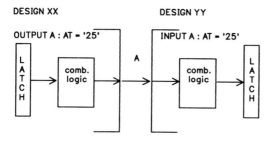

Figure 103. Physical attribute "AT"

3.2.4 The Macro Level

The Design Language supports a macro library, which can be accessed at compilation time. Macros are typically used to specify hardware elements which cannot or or should not be synthesized. Supporting a hierarchical design methodology the macros are also used to define user functions. Examples for macro invocations were shown in Figure 102.

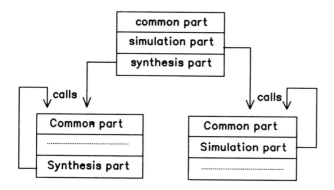

Figure 104. Macro hierarchy

As indicated before, the Design Language is used for 2 purposes: It serves as the input for the LSS processor, which generates the actual, (technology dependant) logic. In addition, it describes the structure to be implemented to the Simulator for design verification purposes (see "3.7 Logic Simulation" on page 190).

The macro language allows the separation of simulation and synthesis information. Depending on whether the compiler runs for simulation code generation or synthesis input data generation, one of the two macro parts are expanded (Figure 104). Macros can be nested to any depth. Care must be taken to guarantee consistency between the different macro parts. The macro facility is the main feature that makes the Design Language suitable for large designs where synthesizeable logic and critical path logic have to be implemented.

A macro definition contains input and output declarations and statements, or a PLAs like design. It might also contain other macros.

In addition to signals, **registers and arrays** can be declared. These are required to describe the behavior of storing elements like latches and memories. Unbalanced IF statements (i.e. having no ELSE path), which are not allowed for designs, can be used for these elements.

The IBM REXX [IBM4] language was used as a macro expansion control language. Macros can be called with certain parameters. A typical parameter is "WIDTH" (Figure 102) which refers to the bit-width of the latch. For the simulation description the latch is WIDTH bits wide, whereas for the synthesis description WIDTH latches are generated because the technology SRL (shift register latch) is only one bit wide.

3.2.5 The System Level

On the System Level the Design Language is used to describe boards, cards or chips which consist of several hardware blocks that were specified on the design level. The language only supports the invocation of these blocks and their linkage via signals. If input/output signals of several blocks are the same an implicit connection is made by the compiler, i.e. no extra linkage information needs to be provided on the structure level. No combinational logic statements are allowed on this level. The syntax of the system level looks very much like the macro invocation syntax on the design level (Figure 105).

A system level description is called a "STRUCTURE". Structures can have input and output signals with attributes. Nesting of structures is allowed to any depth. During simulation, signal names have a hierarchical identifier to specify their structure/design/macro name and level.

The structural hierarchy normally reflects a project's hardware packaging structure. Similar to the way designs are generated as input into the LSS processor, also the structure description is used as input to the LSS processor to generate the appropriate technology description for cards and boards. Of course LSS does no logic optimization on this level. However it is guaranteed that the simulation description (i.e. the logical wiring) is identical to the information passed to the physical design process.

3.2.6 Design System Dataflow

Figure 106 shows the dataflow in our Design System. The Design Language compiler consists of two parts: the design level and the system level compiler. For simulation the design level compiler generates segments of S/370 code and the system level compiler links these segments according to the logical wiring information. For the synthesis path LSS input is generated. The technology

```
STRUCTURE
OUTPUT ADDRESS(0..31), DATA(0..31);

STRUCTURE PROCESSOR

BUS_REQUEST....| BUS_REQUEST    |
               | BUS_BUSY       |....BUS_BUSY
               | BUS_GRANT      |....BUS_GRANT_OUT0
               | ADDRESS_BUS    |....ADDRESS(0..31)
               | DATA_BUS       |....DATA(0..31)       ;

DESIGN UNIT1 : BUS_UNIT

BUS_GRANT_OUT0....| BUS_GRANT_IN   |
                  | BUS_GRANT_OUT  |....BUS_GRANT_OUT1
                  | BUS_BUSY       |....BUS_BUSY
                  | BUS_REQUEST    |....BUS_REQUEST
                  | ADDRESS        |....ADDRESS(0..31)
                  | DATA           |....DATA(0..31)    ;

                  "
                  "
                  "

DESIGN UNITN : BUS_UNIT

BUS_GRANT_OUTN-1..| BUS_GRANT_IN   |
                  | BUS_GRANT_OUT  |....BUS_GRANT_OUTN
                  | BUS_BUSY       |....BUS_BUSY
                  | BUS_REQUEST    |....BUS_REQUEST
                  | ADDRESS        |....ADDRESS(0..31)
                  | DATA           |....DATA(0..31)    ;

END STRUCTURE
```

Figure 105. Example for the system level

output which is generated by LSS is checked against the design source (it is also possible to check it with mixed level simulation).

3.2.7. Overall Comparison with VHDL

VHDL is the standard hardware design language proposed by the USA Department of Defense [IEEE]. For a comparison with VHDL all our hardware design tools have to be considered. For simulation on the functional or procedural level there is a Behavioral Language (see "3.7.2 Hardware Specification Languages" on page 194). Hardware pieces can be described in their behavior using this language. The system level compiler is able to link together pieces designed in the Design Language and/or the Behavioral Language.

The scope of VHDL: architectural design, behavioral design, structural design are covered by our languages. Some features which are implicit and fixed in

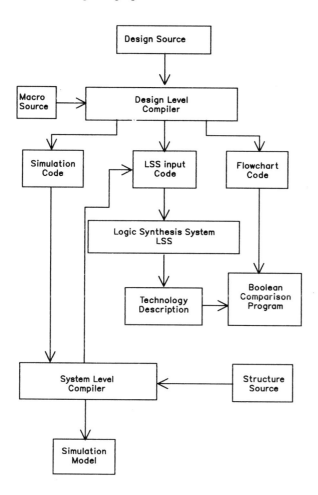

Figure 106. Design system dataflow

the Design Language can be user-defined in VHDL. Examples are the values and value types of signals and the bus resolution function which is totally user-specified in VHDL whereas the Design Language has only three possible algorithms. VHDL is a language where many features are user-tailorable but on the other hand this means that the user MUST specify the features he really wants. This makes VHDL much more complicated than the Design Language. In this sense the Design Language is like a handy subset of VHDL. Each project using VHDL with "real" hardware designers will have to tailor its own subset.

Logic Synthesis is an open question concerning VHDL. The synthesis concepts are clear and distinguished in our language. For pure behavioral descriptions (NOT being synthesized) there is a separate language, stressing the fact that this is a different medium of description. VHDL allows all possible mixes of description style in one document. For projects applying synthesis here again a limitation to a proper subset of the language's possibilities is vital.

VHDL has certain features which are directed to design management (e.g. configuration bodies). These features are not reflected by the Design Language but left to the underlying operating system and a design data management system.

3.3 Logic Synthesis

Bernhard Kick

3.3.1 Overview

The Logic Synthesis System (LSS), originally developed at IBM's Thomas J. Watson Research Center, is the tool that accepts a technology independent Register-Transfer description of some digital logic and generates a technology specific circuit implementation. This tool has been previously used to synthesize chips for the IBM 3090 mainframe computers employing bipolar technology [DAR2]. The Capitol chip set was the first application of logic synthesis in the development of CMOS chips.

The designs are specified on a (mostly) technology independent level using a hardware description language, the Design Language. The designs are used as simulation models for early functional verification of the design (see "3.7 Logic Simulation" on page 190). The very same designs are "compiled" to a gate level implementation by LSS. This ensures that the gate level implementation (BDL/S level) is functionally equivalent to the Register Transfer description (Design Level).

A complete VLSI chip is too large to be synthesized in one piece. We therefore divide the chip into several partitions (see "3.3.5 Partitioned Synthesis" on page 167) and synthesize each partition separately. After linking the partitions, timing requirements and global fanout data are generated automatically. These are used to guide the optimization process during subsequent synthesis runs.

3.3.2 Logic Synthesis Methodology

An overview of the data flow through the synthesis system is given in Figure 107.

Logic designers specify their logic at the Register-Transfer-Level using the Design Language. The Partitions are compiled into a network of boxes, connected by signals. Boxes correspond to the operators in the Design Language, while signals correspond to variables. No optimizations are performed by the compiler, this is left to the synthesis system.

LSS processes each partition, producing a technology specific gate level implementation. This is done by applying a sequence of transforms to the logic con-

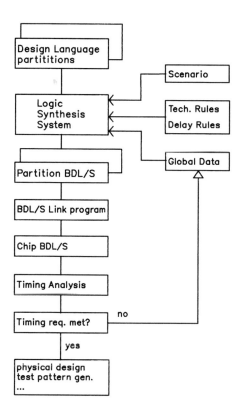

Figure 107. Logic synthesis methodology

tained in this partition. The list of transformations to be applied is specified in the scenario file. Optimization decisions are guided by technology rules and delay rules, describing specific characteristics of the target technology. During the first pass through LSS, global timing and fanout data are not available since these depend on the implementation of the other partitions that have yet to be synthesized. Hence we assume defaults about drive capability and fan-out of partition inputs and outputs, also we do not perform timing correction. The result of synthesis is a technology specific gate level implementation for each partition. The description language used at this level is BDL/S, an IBM internal standard language [MAIS].

The partitions are then linked, resulting in a BDL/S description of a whole chip. Linking is a relatively straight-forward process of identifying chip I/Os, and renaming the partition I/Os such that the partitions become properly connected.

Timing analysis is performed on the chip BDL/S to generate timing data for partition inputs and outputs, and to collect information about global fanout. Usually a chip does not meet timing and global fanout requirements after the first pass. We therefore re-synthesize the partitions using the global data to guide timing correction and global fanout correction transforms. This results in an implementation optimized according to the timing requirements. To get more accurate global data, this process is iterated a few times.

Assuming the chip meets the timing requirements, logic design is completed. The BDL/S is released to physical design where placement, wiring, mask data generation and test data generation are performed (see Part 6). Otherwise the designer has to change his description and to re-synthesize his design.

3.3.3 LSS Overview

The LSS Processor reads several types of input data:

- Logic input generated by the Design Language compiler. This is the primary input to LSS, describing the logic as a network of boxes connected by signals. The file contains the list of inputs and outputs, the list of gates (boxes), and for each gate their input and output signals.
- Scenario file. The scenario specifies the sequence of transforms LSS should apply to the logic to generate a logic gate level implementation. The scenario basically is a sequential list of the transforms that should run. There are also constructs that allow repetitive execution of transforms.
- Technology data files. These file describe characteristics of the target technology, such as the available book set, and delay equations of each book. The information is used to enable transforms to make technology specific optimization decisions.
- Global data. These files describe characteristics of global nets i.e. nets that interconnect partitions. This information is used to guide global fanout correction and timing correction.

LSS manipulates the input description by applying local transforms. Transforms change the logic on several levels of description:

- Design Language level
- Decode/Select level
- AND/OR level
- NAND level
- Technology level

The first level is the description produced by the Design Language compiler, where boxes correspond directly to constructs of the language; the last level is the technology level, where each box directly corresponds to a technology primitive. At each level transforms are applied to simplify the design, and to move it to the next lower level.

Design Language level: At this level transforms deal with operators produced by the language compiler. From logic expressions the compiler generates 2-way AND/OR fanin trees. IF and CASE statements are compiled into CASE boxes. Transforms compress fanin trees into single boxes. CASE boxes

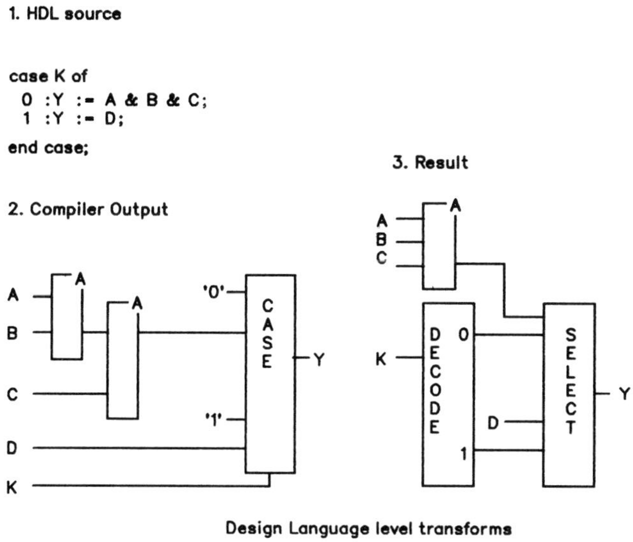

Logic Synthesis

1. HDL source

```
case K of
  0 : Y := A & B & C;
  1 : Y := D;
end case;
```

2. Compiler Output

3. Result

Design Language level transforms

Figure 108. Design language level transforms

are expanded to DECODE/SELECT structures, moving the description to the decode/select level (See Figure 108).

Decode/select level: At this point "high level" operators, such as selectors, decoders, parity generators, adders, etc. are processed. There are transforms to detect and generate decoders from random logic. Decoder optimization transforms use the orthogonality of the decoder outputs to simplify the logic. Figure 109. shows how a small example is simplified by a sequence of transforms. Signal G gates all data inputs to the selector. Therefore it can be factored through the selector, replacing 4 ANDs at the selector input by one at the selector output. Since decode outputs 0 and 3 gate the same signal they can be combined, decreasing the size of the selector. Finally, orthogonality is used to minimize the number of decoder outputs used. ("decode 0 or decode 3" is equivalent to "not(decode 1 or decode 2)"). This results in a much simpler expansion of the decoder since only two of its outputs are used.

AND/OR level: AND/OR transformations perform simple boolean optimizations such as constant propagation, and common term elimination. Translation to NANDs is performed which allows transforms to apply to a network whose nodes are only NANDs and registers.

NAND level: Besides local NAND transforms some global optimizations are applied at this level [DAR1]. Redundancy removal transforms eliminate connections that can be replaced by a constant, thereby eliminating testability problems [BRAN]. Other transforms based on dataflow analysis reduce connection count and try to move controlling signals forward in the logic [TREV].

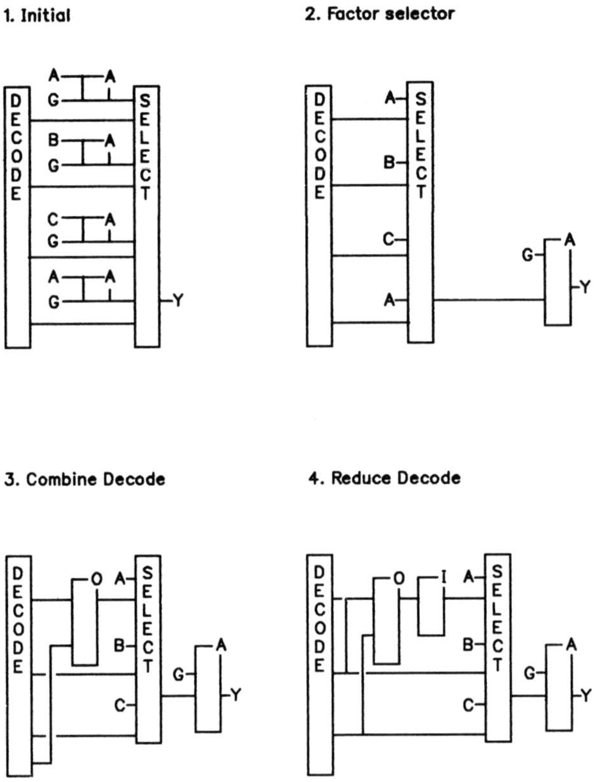

1. Initial   **2. Factor selector**

3. Combine Decode   **4. Reduce Decode**

Figure 109. Decoder/selector transforms

Technology Level: At the technology level of description transforms are applied to enforce technology restrictions, and to make good use of available technology primitives. Fanin constraints are enforced by factoring common inputs to parallel gates, and by generating fan-in trees. Figure 110 shows an example of this. Both gates exceed the fan-in limit of four. The factor transform pulls out the common inputs A and B thereby reducing the fan-in of the gates. Fanin to the gate feeding Y is then corrected by another transform which adds a fan-in extender gate.

Fanout correction is done by inserting buffers, or copying the source gate. Inverter reduction takes advantage of complementary outputs (at registers, off chip drivers etc), and pushes inverters into source or sink boxes. Complex technology primitives, such as AND-OR combinations and XORs, are generated where they reduce cell count. Figure 111 shows an example of inverter reduction. First it is recognized that an inverter can be saved by inverting the inputs to the AI, changing its function to an OR. This results in another opportunity to save an inverter by pushing it left into the AND.

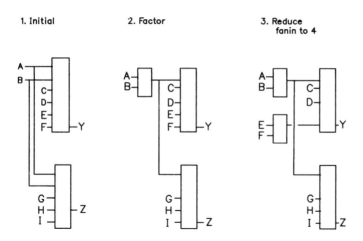

Figure 110. Fanin correction

Timing Correction: Up to this point transforms were aimed to reduce cell count and connections, while timing considerations were of secondary importance. Timing correction focuses on critical paths, aiming to reduce path delay even at the cost of increasing cell count, or lengthening other, less critical paths. Since timing correction is done at the technology level, the propagation delay of all gates is precisely known. At this point no placement information is available. Since the wiring length is a significant contribution to the overall circuit delay, wiring capacities are estimated, based on the fanout of a signal, and are verified at a later step.

Very often circuit count (cell count) and speed are conflicting objectives. To avoid an excessive increase in cell count and/or power consumption the application of the transforms is restricted in several ways:

- Transforms that shorten paths at the cost of increased cell count and/or power consumption apply only to paths that do not meet the timing requirements (critical paths). Often only a few critical paths cause the logic to miss the timing requirements. By applying costly transforms to critical paths only, the cell count increases only where necessary.

- If a transform can shorten a path only at the cost of lengthening another, it checks that the lengthened path remains less critical than the shortened path. In other words, a transform is not allowed to solve one problem by creating another more severe one.

- Transforms increasing cell count apply only if they improve timing by a certain minimum amount. This minimum improvement is passed as a parameter, controlling the size/delay trade-off made by the transforms. Transforms are applied several times, starting with a high threshold, decreasing it in subsequent applications. Hence transforms first apply at places where they most effectively improve timing. This tends to eliminate

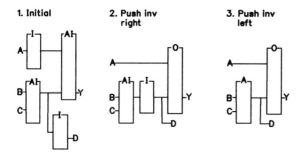

Figure 111. Inverter reduction example

less efficient (in terms of delay) but equally costly (in terms of power and size) changes to the logic.

The basic structure of the timing correction transforms is shown in Figure 112: If, for a certain transform, it is easy to predict the effects on delay then the overhead of undoing the transform can be avoided by adding the improvement test to the first IF in the algorithm outlined above.

Figure 113 shows how timing correction improves a fan-in tree. Fanin ordering moves the critical signal A forward, swapping it with E to maintain the fanin limit of 4. If signal A is still late, fanin reduction splits the 4-way AND into a 3-way AND feeding a 2-way AND. Note that fanin ordering does not cost cells, and can be applied throughout the logic. However, fanin reduction does cost cells. It will apply only if A is critical, and if splitting the AND does improve delay by a minimum amount.

Other timing correction transforms perform fanout ordering and reduction. They move critical sinks closer to the source in repowering trees, replace heavily loaded gates by high-power versions, and reduce the fanout of critical signals by adding buffers.

3.3.4 Technology Information

To decide whether the application of a transform is desirable, most transforms need access to some kind of information about the target technology. This is obviously true at the technology level where transforms by definition have to produce logic conforming to technology rules. It is perhaps not so obvious at the higher, technology independent levels where technology information is used to predict the benefits of transform applications. On the other hand one wants to keep the transforms as technology independent as possible, as to allow for easy adaptation of the synthesis system to different target technologies.

LSS addresses this problem by keeping all technology dependent information external to the transforms, letting the transform access the information as needed to guide optimization decisions. Most technology specific information is

```
Timing Correction Transform:
   for all gates G in the logic do
      if  G with its neighbor(s) forms a logic
          pattern that we can transform
         and
          G is on a critical path
      then
         apply the transform
      end if
      if timing at G has not improved by
         a certain threshold
      then
         undo the transform
      end if
   end for
end Timing Correction Transform
```

Figure 112. Timing correction transform

stored in the form of tables. Thus technology information is concentrated in one place, minimizing the updating effort when synthesizing to a different target technology. Transforms retrieve information from the technology tables through table access routines, so-called user exits. User exits are also used to represent technology information that is not easily expressed in the form of tables.

Information includes for each technology primitive (gates, etc.)

- Technology specific name of the gate (G501, G461, etc.)
- Function of the gate. (AND, OR, REGISTER, DECODE etc.)
- Cell size
- Power consumption
- The minimum and maximum pin configuration indicating what pins must be connected and what pins may be unconnected.
- Information about complementary output pins.
- The pin capacity of each input pin.
- The drive capability of each output pin.
- Delay equations for the fanout dependent propagation delay.
- BDL/S related information (name of pins etc.)

(In IBM design system terminology, the term "pin" refers to the input or output of a circuit, gate or box. This dates back to times, when the task of a design system consisted of wiring discrete circuits on a card or board. The term pin continues to be used, although nearly all circuit I/O's interconnect to other circuits on the same chip).

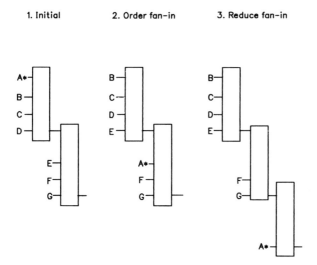

Figure 113. Fanin ordering and reduction

An example of a transform that heavily relies on technology information but that has been made technology generic by means of user exits is the fanout correction transform. The algorithm is outlined in Figure 114. FANOUT loops over all output pins of all gates, correcting fanout violations by inserting buffers, or by copying the source gate. The decision where, and what form of fanout correction to apply is completely under control of user exits. The drive capability of a pin and the pin's actual load are determined by user exits 'load' and 'drive' respectively. 'Drive' determines the drive capability of a pin, either by taking it from the technology tables, or by assuming a reasonable default value if the gate is not a technology primitive. The load capacity is the sum of the pin capacities connected to the signal plus a fan-out and technology dependent wiring capacity. The transform first determines whether it is allowed to apply default fanout correction to a pin (statements "if CanPower (P)..."until "else") leaving to CanPower the decision which pins are allowed to repower. If default fanout correction is not allowed, FANOUT calls a user exit UserFanout (the "else" part in Figure 114) to perform user defined fanout correction. (e.g. this was done to do a special form of repowering in clock distribution logic.) If normal fanout correction is allowed, (CanPower (P) = TRUE) then the decision whether to insert buffers or to copy the source gate is left to another user exit. "MayCopy" (see again Figure 114) is an example of a user exit representing technology information that is not easily expressed in the form of a table.

While the implementation of the user exits is of course technology dependent, the interface to the transforms is independent of the technology. User exits provide an easy-to-use and stable interface, useful at all levels of description, and independent of the target technology. The interface remains stable even

```
procedure FANOUT:
  for all boxes B
    for all outpins P of B
      if load(p) > drive(p) then
        /* fanout limit exceeded */
        if CanPower(P) then
          /* normal fanout correction */
          if MayCopy(B,P) then
            /* copy the source gate */
            call ParallelFanout(B,P)
          else
            /* add buffers */
            call SerialFanout(B,P)
          end if
        else
          /* pin needs special treatment */
          call UserFanout(B,P)
        end if
      end if
    end for
  end for
end procedure
```

Figure 114. Outline of fanout correction transform

when the internal format of the tables changes. If characteristics of the technology change, only technology tables and user exits have to be updated; the transforms themselves remain unaffected.

3.3.5 Partitioned Synthesis

A VLSI chip containing 30,000 gates is too big to be synthesized as one piece. Due to memory space and run time limitations, a single chip is divided into several interconnected partitions. Partitioning hardware has advantages that parallel those found in modular software designs. First, it allows several designers to work on a chip concurrently. In addition, changes require only the affected partitions to be re-synthesized instead of the whole chip (see "3.5 Timing Analysis and Verification" on page 177). The same partitions were used for floor planning and placement in physical design (see "6.2.2 Partitioning and Floorplanning" on page 285).

There are some problems with partitioning, too. While most of the synthesis tasks can be performed independent of other partitions, there are two tasks that require knowledge about the partition's interconnections and the implementation of other partitions: fanout correction of global nets, and timing correction.

These problems are addressed in LSS by providing a global data table, holding information about global nets, i.e. partition inputs and outputs. Global data are extracted by the timing analysis program which processes the whole chip (see "3.5 Timing Analysis and Verification" on page 177).

Global data include for each global net:

- Type of source gate
- Number of sink partitions, and load in each sink partition
- Actual arrival time
- Requested arrival time

Global data also include the cycle time, and the latch- and trigger-times of all clocks.

Data like the type of source gate, and the number of sink partitions are used to guide global fanout correction. Since it is not allowed to add a partition output (the global interconnections are determined by the designer, not by the synthesis system), fanout violations of global nets usually cannot be corrected by repowering the source gate. For that reason global fanout correction aims to reduce the load in each sink partition. We adopted the following strategy to ensure proper fanout of global nets whose fanout limit is exceeded:

- In the source partition we insert a buffer thereby increasing the drive capacity of the output.
- In the sink partitions we allow the global net to have a maximum fanout drive capability of source gate/number of sink partitions, which ensures that the total fanout of the net is below the limit.
- If the global net is timing critical then sinks that are not timing critical are forced to use a buffer to minimize the global fanout of the net.

Figure 115 shows how global fanout correction works on a small example. The source gate of X has a fanout limit of 6, the total fanout is 9, and there are 2 sink partitions (B, C). To enforce the fanout limit, we first reduce the fanout to $6/2 = 3$ in each sink partition. Since X is timing critical, a buffer is added in partition C and the sinks in partition B are reordered which reduces the global fanout to 3.

Timing correction is based on knowledge about critical paths in the logic. To determine those paths, actual arrival times are propagated forward through the logic starting at partition inputs and latch outputs, and required times are propagated backwards starting at partition outputs and latch inputs. The global data provide enough information to perform timing correction in a partition. Timing at latch I/Os is derived from the clock timings, which are part of the global data. A latch output arrives at the trigger time of its clock (plus a load dependent latch internal delay). A latch input is requested to become stable before the latch time of the clock (minus a latch dependent setup delay). Timing of partition I/Os is taken from the global data. At partition inputs also

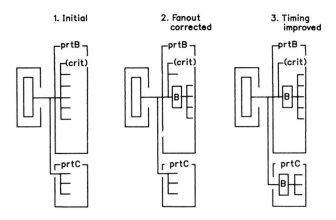

Figure 115. Global fanout correction (max. fanout = 6)

the type of source gate is known, therefore the effects of changing the load of a partition input can be exactly determined.

After the first pass through LSS, global data are only a crude approximation to the final data, because there is no timing correction in this pass (due to lack of timing data). In the second pass, fanout and timing correction is performed based on the global data from the first pass. Since the global data are not very accurate in the first place, the optimization result cannot be very accurate either. However, the implementation is now closer to the final one, resulting in global data that more accurately reflect the timing requirements that have to be made for each partition. The process is now repeated until the global data become stable. While we have no theoretical proof about global data convergence, we found this method to be quite effective in practice. In our experience data become stable after 2 to 4 iterations.

For a description of the global timing correction procedure, see Figure 121.

3.3.6 Synthesis Experience

Early in the design cycle we used logic synthesis to get a prototype implementation for cell count and delay estimates, and to evaluate the quality of the synthesized logic. This early feedback was very important for both the chip designers and the synthesis system designers. Chip designers identified critical parts of their design and were able to change it accordingly. At the same time, synthesis system designers eliminated many LSS inefficiencies pointed out by the hardware designers. Transforms were developed, targeted specifically at these inefficiencies. Especially in the area of timing correction much effort was spent. The modular structure of LSS, the layer of database access functions provided, and the ability to interactively apply transforms and evaluate the results, turned out to be the key feature to enable rapid development, debugging, and application of new transforms. In summary, the value of the highly cooperative effort of hardware and software designers can hardly be overestimated; it was the key to the success of logic synthesis in this project.

While cell count usually met the requirements, timing requirements turned out to be more of a problem. LSS was able to satisfy timing requirements on most of the paths, but there remained areas in the design where the synthesized implementation did not meet the tight delay requirements. Compared to an unoptimized implementation, timing correction reduced critical path lengths by typically 20% to 30%, increasing cell count by approximately the same amount. The remaining timing problems had to be fixed manually. This was done by allowing the designer to specify logic directly on the technology level. Thus Design Language descriptions are a mix of register transfer level, and low level, technology specific descriptions. LSS is still used to synthesize the technology independent parts of the description, while passing low level logic unchanged. Low level design is much more tedious and time consuming than higher level design, but it gives the designer immediate control of the implementation.

While LSS could not solve all timing problems, it was still of great value as it allowed the designers to concentrate on the few critical places, leaving the rest of the chip to logic synthesis. It turned out that often it was data flow logic that LSS did not handle very well. This type of logic is characterized by its very regular structures. Generating efficient implementations from these structures seems to require global planning, which does not match well with the LSS paradigm of local transforms. For instance the ALU logic on the CPU chip had to be designed on the technology level. Overall less than 20% of the logic, mostly dataflow logic, was designed this way.

To synthesize the final designs in a production environment, we used LSS as a fully automatic tool with a fixed sequence of transforms. Synthesis runs were made over-night on a batch machine. The CPU chip was synthesized as 30 partitions of roughly equal size (about 1000 gates per partition). It could be synthesized in about 5 hours of CPU time on an IBM 3081. This time includes one complete LSS run with timing correction, the partition link process, and timing analysis. Because a few iterations were needed to get accurate global data, it took 3 to 5 days to completely synthesize one chip. Small to medium size design changes could usually be synthesized within one day, using the existing global data. The other chips were of similar size with similar run times.

3.4 Logic Synthesis Design Experience

Hans Kriese

3.4.1 Overview

The Capitol chipset processor design consists of old design pieces and new parts. The old pieces are a carry over from the IBM 4361 and IBM 9370-90 design. They were represented in a graphical form (at a bipolar technology circuit level) from where BDL/S is generated automatically. These old pieces had to be manually transposed ("mapped") from bipolar into CMOS technology. The new parts had to be designed with the new technology in mind. CMOS offered such a variety of basic circuits that only a logic synthesis program would be able to really exploit the possibilities. In addition the details of the new technology were not defined at project start, so that the design either could not start (using the "old" graphic entry at circuit level) or it had to be technology independent. We therefore choose the technology independent Design Language described in "3.2 Hardware Design Language" on page 148, which requires LSS to generate BDL/S automatically. This choice also resulted in "mapping" the (technology dependent) graphic description of the existing old design part into the Design Language.

3.4.2 The Design System

The Design System (Figure 116) offered two design entry alternatives: either using the conventional way of graphically specifying the structure of technology circuits to be translated to BDL/S for chip manufacturing or specifying the design implementation (logical function) with the design language to be compiled into LSS input. LSS then would take that description, apply some algorithms (called "transforms") for boolean minimization, floor space and connection reduction that are technology independent and others that are technology dependent and try to reduce again floor space and connections, and then employ algorithms that try to optimize the delays according to given constraints "3.3 Logic Synthesis" on page 158. Output is again BDL/S for chip manufacturing. Design Verification may take place either starting from the design language or from BDL/S or from both mixed ("3.7 Logic Simulation" on page 190).

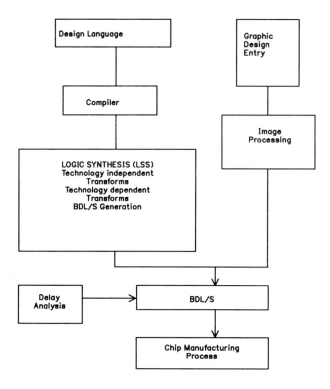

Figure 116. Design system

3.4.3 Challenges in Using LSS

The use of silicon compilation is still evolving. There are similarities to the introduction of high level languages replacing assembler language in the 1960's and 1970's. The IBM approach to silicon compilation, LSS has been used successfully in large computer designs where the primary goal was to save design time. The design pieces had timing constraints, but not so much floor space constraints.

When designing in a CMOS technology, the trend is to put as much as possible on one (or a few) chips. This leads primarily to floor space constraints, so that LSS, when it was tuned from bipolar technology support to CMOS technology support, primarily worked towards saving space and connections.

The Capitol chip set design had both to fit on a certain area and had significant timing constraints. The first challenge was the partitioning into pieces so that LSS could not see the whole design at a time about 60.000 2-way equivalent NAND gates). This created global optimization problems.

The second challenge was the use of a new methodology that was still in the process of being developed (and debugged) while being used as a design tool.

Logic Synthesis Design Experience

At project start the transforms for the technology related optimization were not yet complete, which resulted in unrealistic space and time figures.

The third challenge was the fact that the logic design was not new, but was mostly "mapped" from an existing bipolar design with a moderate (700 circuit/chip) integration level. When evaluating the possibilities of the chosen technology, the most time critical path was calculated resulting in a certain maximum delay figure, thereby assuming that the control logic could be arranged accordingly. In the mapping process, however, where designers described the design in our design language based on the graphic representation of the old technology circuitry, all constraints regarding the multiplexing and decoding of signals crossing the old (and many) chip boundaries (limited pin count), were also described in the design language. The old chips forming the natural design pieces to break the whole design into, these constructs could not be seen by LSS when processing one design piece at a time. Thus the logic generated piece by piece and finally put together still had all these constructs in it and did not reach the timing goal initially set by calculating the critical data path. The goal for the data path was reached, but the control logic, that was mapped instead of rearranged, became the gating factor.

3.4.4 Delay Optimization and the Use of LSS

In the beginning LSS would process the design (piece by piece) without any external timing information. Then the delays of the paths of the whole design (composed from the design pieces that were synthesized one at a time) through the logic were calculated by the timing analyzer (see "3.5 Timing Analysis and Verification" on page 177). The figures were gathered in reports on delays and also on cell counts for the various circuits. Paths through the whole design were the primary feedback for the designers on how LSS had done its job. This was rather detailed information. The figures were kept in listings (delay per net on design piece boundary) together with the figures that were obtained by calculating the required maximum delays that would allow a certain machine cycle time. These lists were fed back to LSS for an iteration run. Then the delay calculation was repeated, the actual (and desired) delay figures were fed back to LSS again etc.

The result of the iteration process was (as anticipated) that the "long" or "critical" paths became shorter and the uncritical paths became longer, but usually under the critical limit (see Figure 117). This process was converging, so that after two iterations it turned out to be stable. The delay optimizing process within LSS (the transforms) takes the figures and according to another (LSS internal) delay calculator would either rearrange the logic or insert power-up circuits (see "3.5 Timing Analysis and Verification" on page 177).

With this method we even achieved a good result (with only 6 manually inserted circuits) for a design using multiple clocks. This tends to make delay tuning very difficult. However, our timing analyzer provided detailed informa-

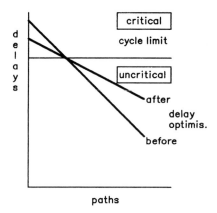

Figure 117. Delay distribution

tion on the delays of the long paths dependent on the gating clock, which we then could feed back to LSS.

3.4.5 Results and Designers' Echo

Now what were the results? Considering the challenges, the net result was very positive. We achieved excellent floor space results. We had cases where for smaller pieces designers had made manual layouts of the logic, and where LSS, "enriched" with our technology specific transforms, made it significantly better. However, we generally exceeded the maximum delay figure that was calculated for a manual design by about 15%.

These results were obtained through constant improvement and development, especially in the area of delay optimization as described above.

The feedback we got from our designers helped to detect weaknesses in the transforms or the handling, and usually corrections were made on the spot. The permanent changes, however, sometimes resulted in inferior figures for areas which designers considered already finished. In those cases the feedback was not restricted to a pure technical communication. Finally LSS delivered stable results.

We had logic designers of different background: some were only used to graphical design entry at circuit level, some had already used a language to describe design pieces for simulation, and some were totally new. Let us start with the individual designer's echo.

The "new" designers generally were satisfied by the results they got from LSS. They, however, were designing the more peripheral portions that were not so time critical. They liked the whole design system. They did not need much education and did all work by themselves, including the LSS runs for their design pieces.

The "graphically" oriented designers were not satisfied with the results from LSS. They had the central and really time critical parts. They liked the easy design entry via design language, but they complained about the lack of pictural representation of the design. Especially bad was in their opinion the graphical output of LSS that arranges logic based on algorithms so that the layout really had no similarity of what they would have entered graphically themselves. Another point was that LSS did not show them incremental improvements to be compared with previous results, but always (in case of a logical change or a change in the transforms) after a new LSS run the design looked again totally different. This made it also very cumbersome to insert changes into the output of LSS, comparable to making fixes in the object code instead of in a Higher Level Language source. When tedious timing corrections with maybe some manual insertions of speed-up drivers were done, a source change and rerun of synthesis might change them, requiring to redo the timing correction, potentially quite different than before. This group did not like to run LSS by themselves, since they were "busy designing".

Other designers with both, graphical and language entry experience said, they could not have finished their design piece in such a short time if they would have done it using graphic input. They also were content with the results.

One common thing could be observed with all designers: they considered synthesis as being a black box monster swallowing their design, and they had no influence of what would be produced. They understood single steps of LSS, they also made proposals how to improve things, but they still felt the whole process was out of their control.

All designers liked the feedback that was formed by the timing analyzer consisting of the delay paths and cell counts for the whole design (or for pieces).

The general echo of the design group was that LSS didn't do the job they had expected. It was not sufficient to reach a maximum delay 15% above a manual design. They claimed that with a graphical entry at the technology circuit level they could have reached the delay goal and also the floor space goal.

Management (e.g. the author of this book) believed in a shorter product development cycle, mostly because logic design verification (see "3.6 Logic Design Verification" on page 184) could start significantly earlier. The assumed schedule gain of 3 - 6 months was probably partially offset because the LSS tools had previously been used for bipolar designs only, and were less mature than originally assumed in working with CMOS.

3.4.6 Discussion

The discussion of LSS as design method was soon augmented by a discussion about graphic design entry versus list language. The latter usually requires the synthesis process, the first is generally (or historically) translated into an exact structural description of the picture ("blocks and nets"), in our case resulting in technology related BDL/S. This process may include some "expansions" of complex functions into several less complex functions, but not synthesis itself. The graphic versus language discussion has a more emotional aspect of what is easier to understand (read) by the individuum, which will be neglected.

It has been shown that design entry in language is faster. Together with LSS it allows an earlier feedback on the physical aspects of the design (floor space, delay, power consumption, etc.). The computer processing time, however, is much longer for LSS than for a graphic translation to BDL/S. As a consequence a late small change has a much longer turnaround time through source language and LSS than through graphic update and translation. (This could be less important, when the design pieces handled by LSS at a time are rather small, but LSS is conceptually to be applied to the whole, limited only by the patience of the designer to wait for results).

While dataflow logic is highly regular or structured, and therefore easily to be entered in graphical form, it is a tedious task to do the same for control logic, where LSS would probably make even better logic than a designer. In addition the entry can be done much faster in design language due to constructs like "CASE", "IF THEN ELSE", "PLA" (see "3.2 Hardware Design Language" on page 148).

Another point of discussion was "technology independent" design. When the target technology of a design is not yet defined, it appears advantageous to describe the function in terms that allow a fast implementation into the one or the other technology. The design language offers that capability naturally. A graphic design entry has to make use of higher level function blocks (e.g. multiplexer, registers, decoders, etc.) that need to be expanded into the technology circuits in a synthesis - like manner, or the expansion had to be done by hand. LSS can be adapted to a certain type of technology (e.g. CMOS) rather quickly, but the tuning will take some time according to how close to the optimum it should be.

3.4.7 Conclusions

It appears that both methods, synthesis (via LSS) and graphic entry have advantages and disadvantages. It may be advantageous to support both ways for the future to allow the designer to work efficiently using the design tool he chooses. It appears that for a VLSI design with critical requirements for both space and delay, the pure application of LSS is not the best approach to achieve 100% of what can be achieved.

Logic Synthesis Design Experience

The use of mature LSS tools will result in shorter project development schedules. Currently LSS is restricted to logical manipulations. It would be interesting to have additional capabilities like placement and wiring support based on algorithms and also a quick and efficient interactive user interface. Also, an LSS synthesis package should never be a black box, but should permit the logical designer to enter improvements as they become feasible.

3.5 Timing Analysis and Verification

Siegfried Heinkele

3.5.1 Overview

The new design methodology of the Capitol chip set generated new requirements for timing verification. The size of the design required to split the chips into several partitions, that were synthesized independent of each other using LSS. With the LSS internal delay calculator, timing requirements could be determined as long as the paths were totally within the partition. However the large chips include paths which spread over several partitions. Therefore an additional tool, the "Timing Analyzer", was required, which could calculate timings for paths which spread over several partitions. The knowledge of the slack, defined as the difference of the actual and required arrival time of a net, will not be sufficient. This is because each partition should correct only a part of the overall slack, such that the total path delay meets the requirement.

The generation of these timing requirements for partition boundary nets was one of the main reasons for the development of a new Timing Analyzer. It was written to include additional capabilities, such as Multiple cycle paths, Multiple clock designs, and Wiring capacitance estimate using floorplan data.

3.5.2 Delay Equations

Besides stage counting, the Timing Analyzer is capable to handle various types of circuit delay equations. However, unsophisticated delay equations do not adequately reflect CMOS delay behavior. The delay calculator approach described in "6.4 Delay Calculator and Timing Analysis" on page 296 yielded a model for the circuit behaviour that proved to be adequate.

The following is an example for a non-inverting circuit:

```
Trise = (K1  + K2*C)*Tr  + K3*C**2  + K4*C  + K5
Tfall = (K6  + K7*C)*Tf  + K8*C**2  + K9*C  + K10
Ton   = (K11 + K12*C)*Tr + K13*C**2 + K14*C + K15
Toff  = (K16 + K17*C)*Tf + K18*C**2 + K19*C + K20
  C : capacitive circuit load in pF
  Tf: slope of the up-going edge at the input
  Tr: slope of the down-going edge at the input
  Tx: Tr or Tf
```

The coefficients K1...K20 are calculated with a least square fit method.

These delay equations define the delay and slope from an input to an output pin, using the design dependent parameters C and Tx. The Tx values must be declared for the primary inputs. For all other nets it will be calculated with the above formula.

For the same input-to-output pinpair we allow up to three sets of delay equations, a worst case, a best case, and a nominal case equation set, reflecting the various conditions of temperature, voltage and process parameters. An option in the Timing Analyzer defines which case is to be used.

Additional flags are required to describe the input-output-pin path correctly :

- The **Inversion flag** tells, whether the path is inverting or not. If this is not clear like in the case of an exclusive OR, both an inverting and non-inverting path need to be defined, where normally the different paths will also have different delay equations.

  A **Trigger flag** tells, whether the path is combinatorial or triggered, and whether the delay path (from the previous circuit) should terminate and a new path should start.

3.5.3 Capacitance Estimate

The total capacitive load of a net can be seen as the sum of the wiring capacitance and the pin capacitance of the sink circuits. While the pin capacitance is independent of the design and can therefore be declared in a table for each circuit pin type, the wiring capacitance estimate is more difficult, since it depends on geometrical data, which are not available in the early design phase. In this phase the logic is described in a high level language, (see "3.2 Hardware Design Language" on page 148), and timing optimization occurs on the gate level, at a time, where information about placement and wiring is unknown.

This forces to use some simplifications. It was assumed that any net distributes with a fixed percentage to the first and second metal layer, and that the geometrical layout is the same for each net. This enables to estimate the wiring capacitance as the product of the pure netlength and a technology dependent constant of the dimension pF/mm. For off-chip nets, the capacitance was estimated by other means, but the same type of delay equation could be used with parameters Tx as input risetime/falltime and C as connector pin capacitance. The only difference to the on-chip delays lies in the interpretation of the output risetime/falltime : while this is of the dimension ns (nanoseconds) for on-chip nets, off-chip nets are assigned a slew rate, giving the slope in the dimension ns/V rather than ns.

With these assumptions just the netlength remains to be estimated. The usual method is to estimate the netlength as a function of the fanout, or more precisely, as a function of the number of connections, (regarding the abnormal case of dotting in CMOS technology). The estimate can either be taken from a table, using the fanout as index, or it can be calculated with the formula:

```
netlength  =   K1  +  K2 * fanout
```

where K1 and K2 are parameters to the Timing Analyzer.

This method is adequate for small designs or chips. However timing problems after wiring will arise, if it is applied to the 12,7 mm Capitol chips. Our design method aimed to avoid or at least reduce this problem. The use of partitions facilitates a floorplan at an early design stage, where timing problems can be checked and solved. With the knowledge of the final partition placement on the chip, and with the knowledge of final driver/receiver placement, almost all nets with an excessive netlength can be captured. We still applied the poor fanout dependent estimate for partition local nets, but applied a routing estimate for partition crossing nets. This was done by taking the center location of the partition as the connection point for all global nets to and from this partition, and applying a minimum spanning algorithm to estimate the final netlength. Since physical design (see "6.2 Hierarchical Physical Design" on page 285) used the same method (i.e. the partitions were wired independent of each other, and partition crossing nets were wired separately), the estimate corresponded reasonably well to the actual results.

The quality of netlength estimates is shown in two examples, the CPU chip and the MMU chip. Both chips are of a size of 12.7x12.7mm. While the CPU chip contains only one small array, the Memory Management chip contains four of them, occupying approximately one half of the chip area. This required further attention for the estimate.

Figure 118 shows the deviation of the estimated netlength from the actual netlength after wiring for the CPU chip employing the 2 different approaches: Using floor plan data in addition to fanout data, and using fanout data only. The class# gives the difference of the real and estimated netlength in mm, i.e. positive values show nets estimated too short, negative values show nets estimated too long.

This picture shows that most of the run-aways could be captured. The remaining run-aways have either been partition local nets, or they have been very long nets of more than 40mm netlength, where the relative error is not large. As it turned out, wiring had no influence on the estimated CPU chip cycle time.

```
          floorplan estimate        fanout estimate

          class#       count        class#       count
          ------       -----        ------       -----
          14.00 :          1        32.00 :          4
          12.00 :          2        30.00 :          7
          10.00 :          4        28.00 :          5
           8.00 :          4        26.00 :         17
           6.00 :         11        24.00 :          5
           4.00 :         37        22.00 :          1
           2.00 :        235        20.00 :          2
           0.00 :       3646        18.00 :          5
          -2.00 :      13500        16.00 :         10
          -4.00 :        966        14.00 :         19
          -6.00 :        221        12.00 :         39
          -8.00 :         87        10.00 :         61
         -10.00 :         13         8.00 :        125
         -12.00 :          3         6.00 :        168
         -14.00 :          1         4.00 :        266
                                     2.00 :        579
                                     0.00 :       3939
                                    -2.00 :      12673
                                    -4.00 :        616
                                    -6.00 :        126
                                    -8.00 :         56
                                   -10.00 :          8
```

Figure 118. Wiring capacitance estimate for the CPU chip

Figure 119 shows the deviation of the estimated netlength from the final netlength after wiring for the MMU chip.

The netlength estimate was less accurate than for the CPU chip. The large difference between estimated and best case cycle time amounted to 5ns. Arrays caused wiring estimate problems, due to the following:

- The pinlocation relative to the chiplocation of the circuit is very essential for arrays, while it can be neglected for all other circuits.
- It is essential, in which direction the array is placed. One of the arrays was ultimately placed as a mirror image.
- The area above an array can not be used for wiring.

Considering these items, the MMU chip can be better estimated.

```
                       Memory Management chip
      floorplan estimate     fanout estimate

      class#     count       class#     count
      ------     -----       ------     -----
       22.00 :     1          26.00 :     1
       20.00 :     2          24.00 :     0
       18.00 :     2          22.00 :     4
       16.00 :     5          20.00 :     7
       14.00 :     6          18.00 :     8
       12.00 :    11          16.00 :    18
       10.00 :    34          14.00 :    27
        8.00 :    68          12.00 :    39
        6.00 :    83          10.00 :    78
        4.00 :   123           8.00 :   173
        2.00 :   390           6.00 :   143
        0.00 :  2321           4.00 :   194
       -2.00 :  5507           2.00 :   424
       -4.00 :   381           0.00 :  2138
       -6.00 :   165          -2.00 :  5619
       -8.00 :    27          -4.00 :   179
      -10.00 :     7          -6.00 :    50
      -12.00 :     1          -8.00 :    23
      -14.00 :     1         -10.00 :     0
                             -12.00 :     2
                             -14.00 :     6
                             -16.00 :     2
```

Figure 119. Wiring capacitance estimate for the MMU chip

3.5.4 Multiple Clock Designs

The problem of multiple clocks is demonstrated with the example shown in Figure 120:

C1, C2, C3, C4 are system clocks, latching some data into registers L1, L2, L3, L4, at the times shown in parentheses. Assuming a cycle of 80 ns, it is immediately clear, that register L1 cannot start at the same time for paths both to register L3 and L4. One-cycle paths require L1 output to start at time 50 for register L4, while it has to start one cycle earlier for register L3. It is not possible to fulfill both requirements at the same time. Several solutions are possible :

- Our Timing Analyzer performs independent path analysis runs for each different sink clock. In this sense primary outputs have to be viewed as clocks, too. This approach seems to be most flexible, because it can always

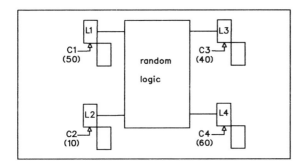

Figure 120. Multiple clock design

be applied. Furthermore for any analysis run, nothing else but the clocking scheme and the requested arrival times of the primary outputs have to be declared. The cycle shifting for the register departure time and primary input arrival time is done by the Timing Analyzer automatically. The criteria whether to shift a signal or register by one cycle is the following : all starting signals are shifted in a way, so that they can reach the latching clock in a one-cycle path.

- LSS has a built-in integrated Timing Analyzer, that applies a more sophisticated approach, performing the analysis not sequentially for each clock, but in parallel in one shot. This requires having an array of delays for each net, rather than changing it several times throughout the analysis. This requires a manual effort to keep the array as small as possible.

- Another approach, the applicability of which mainly depends on the actual design data, is to do everything in one analysis run using a one-system-clock analyzer, where the problem of multiple clocks is solved by manually adjusting the departure times where necessary. Since this solution does not solve the problem in general, it might be necessary to have several analysis runs with different parameters.

3.5.5 Multiple Cycle Paths

Our design had to deal with paths, which were allowed to spread over more than one cycle. Clock gating logic was used to accomplish this. This implies that the start points for many paths cannot be adjusted once for all paths originating from there : there are registers, even triggered with the same clock, some of them requiring the signal in the same cycle, some requiring it one cycle earlier. All these paths have to be analyzed separately, because of different departure times for the start points. In order to avoid these many separate analysis runs, which increase the report size, the Timing Analyzer features a grouping of multiple cycle paths, putting all sinks of multiple paths into the same group, which have the same multiple cycle declarations for the sources.

Timing Analysis and Verification   183

The number of different groups could be better reduced by grouping logically independent cones together. This approach was not implemented due to increases in computer execution time.

3.5.6 Global Timing Correction for Logic Synthesis

Section "3.3.5 Partitioned Synthesis" on page 167 referred to LSS timing optimization. Since a critical path frequently meanders through 2 or more partitions, it is important to perform delay time optimization over several partitions. For this it is necessary to generate global timing data, or "Timing Assertions", a term frequently used in the literature.

We utilize the following definitions:

- Path : a path is a connection between a startpoint and an endpoint in the logic.

- Startpoint : a startpoint is either a primary input (PI) or latch output.

- Endpoint : an endpoint is a point in the logic with a timing test condition. It is either a primary output (PO) or latch input.

- Slack : the slack of a net is the difference of the requested delay and the actual delay. The timing constraints are not met as long as there exists a net with negative slack.

Logic Synthesis will only work on one partition at a time, and will not know the timing constraints of the partition crossing nets, at least as long as timing analysis was not performed on the whole chip after complete synthesis. After timing analysis of the whole chip, critical paths are known and will be reported for each partition. It is obviously clear, that it is not a good method to use the slack of the nets as feedback to Logic Synthesis. The slack must be distributed to all contributing partitions.

Figure 121 shows a small example illustrating the method. The logic path (X-Y-Z) extends over two partitions A and B, and has a required length of 80 ns to fit into a machine cycle. The table shows for each synthesis run the actual path lengths, the required path lengths, and the scale factor used to calculate them. Required path lengths for each partition are generated multiplying actual path lengths by the scale factor. In effect this apportions timing requirements according to actual path length. Long paths are expected to have greater potential savings than short paths. After the first pass, actual total delay exceeds required delay by 40ns. According to actual path length, a timing improvement of 26.7 ns is required in A, and of 13.3 in B. After the 2nd pass partition A has improved by 10ns, but failed to meet its required delay, while partition B improved by 14 ns and is a little faster than required. New delay requirements are calculated based on the new actual delays. As one can see the required delay for A does increase while the required delay for B further decreases. Thus, part of the timing correction burden has moved from

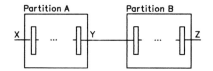

Pass No.		path lengths			scale-factor
		Part. A	Part. B	total	
1	act.	80.0	40.0	120.0	
	req.	53.3	26.7	80.0	80/120
2	act.	70.0	26.0	96.0	
	req.	58.3	21.7	80.0	80/96

Figure 121. Iterated timing correction on a path over two partitions

A (which did not meet the first requirement anyway) to B. In the next pass, LSS will give up on A, but will spend more effort to do timing correction in B.

Timing correction in Logic Synthesis using these assertions turned out to be very successful. The first runs of LSS produced unacceptable cycle times. Applying timing correction reduced the cycle time significantly for both the CPU and the MMU chips. For further reduction, design changes had to be made. Since no timing problems were introduced after physical design, the applied method was judged to be quite successful.

3.6 Logic Design Verification

Hans Kriese

3.6.1 Overview

Until recently, programs that convert the structural description of circuit interconnections into geometrical shapes for manufacturing were frequently the only automatic processes in the whole design flow. All other processes were done manually and therefore subject to errors. "Debugging" the design was a tedious job and took a long time. On the other hand, corrections could easily be implemented due to the low circuit integration level. Today's objective is to assure that every process step in the design flow is automatically done by programs, the only exception being the primary steps at the design level (see "3.2.1 Overview" on page 148). "Design Verification" is the procedure to prove the correctness of the results of these design processes. The logic design verification process is assumed to be fully automated, while the input to it is generally produced by another not fully automated process. Since the processes for verification are different (and they should be independent) from those of the design flow, we have theoretically reduced the probability of undetectable errors by orders of magnitude.

3.6.2 The Concept of Using Logic Simulation

Simulation is a frequently employed design verification method. If we restrict ourselves to verify only the logical behavior, without looking at delays or power or even transient signal levels (binary 0 and 1 only), we call this method "logic simulation".

When the design of the Capitol chip set was started, we found ourselves in the middle of a design methodology evolution. There were excellent results from logic synthesis applications, which encouraged us to use this approach. Special purpose hardware simulation engines were becoming operational but not yet quite ready to be used for the Capitol chip set. We therefore choose to use software simulation systems for our design verification.

Logic simulation as a design verification method has recently demonstrated two trends:

- The level of verification has moved upwards from gates to systems. This is reflected in the size of the model to be simulated as well as in the stimuli used to drive the model.

- The verification period within the design cycle is moving towards the beginning, to the specification phase. Only in case of manual implementation processes (maybe due to late error corrections) another verification cycle has to be added.

We apply logic simulation to our design by building a simulation model for the whole processor and apply processor instruction strings (test cases called BUSTERs, containing "self checking code") directly as test stimuli. This approach allows us to use the same tests both on the simulation model and later on the real hardware (see "7.3.3.1 Sub-Architectural Verification" on page 322). It also allows us to generate the stimuli not per component, but from the description of the processor design ("specification"). The components are tested only in the overall processor system. (Previous projects using the same method have shown that stand-alone simulation of components with special stimuli costs much more time than the system model simulation.) The stimuli for the components are provided by the other components automatically. This approach requires to model also the "outside world" of the processor like I/O interfaces, clock generation ("2.8 Clock Chip" on page 112), service interfaces ("2.11.6 Support Interface" on page 141) etc. to set or serve the signal lines feeding the processor externally or to respond to certain conditions or requests.

In previous projects, design verification used the design representation at the gate level (BDL/S). This was at the end of the design period, since the generation of this BDL/S was "the design". Verification occurred at a late point in the development cycle. The Capitol chip set design utilized logic synthesis. We used primarily the design representation that was the input to synthesis - our Design Language. This permitted design verification early in the development cycle. In some cases we used the BDL/S representation that came out of synthesis for cross-checks.

3.6.3 Modelling Requirements

Logic simulation is based on models of what is to be designed and on models of the "outside world". Both must be fully operational, but the design model has to contain all possible functions (in terms of control logic, dataflow parts etc.) that can be executed by stimuli, while the "outside world" models require only response or triggering on the periphery.

Two types of information about a model must be available: the behavioral information and the structural information. We can consider the input to the design process, the specification, as almost pure behavioural information, while the output of the design process to be used in chip manufacturing is almost pure structural information. The natural flow of design starts with behavioral descriptions and then replaces it more and more by structural descriptions. It is typical for a "top down" logic design to write behavioural models that are the equivalent to a behavioural specification which can be simulated. Adding detail to a design is replacing behaviour by structure of lower level behavioural elements. The "functional level" of a model corresponds to the balance between behavioural and structural information coded into that model. "Functional level" simulation relates to the "system" of models linked together: Some at a high behavioural level, others at a detailed structural level. The "system" of models can vary from a model containing nearly only behavioural elements to be simulated (which we call "high level") to a big structure of circuit models (which we call "low level").

Logic synthesis is a way to generate structure out of behaviour. With its use there is no need to perform "low level" simulation for logic design verification, since this process is fully automated. We can do high level simulation. This is different from a design input system where we have to manually enter the structural information ("bottom up" or "middle out" approach), requiring "low level" simulation. Our simulator is event driven, so that the overhead of scheduling the behavior parts increases drastically for low level simulation. We usually ran into performance or space problems if we wanted to apply a large number of system level test stimuli in low level simulation.

These problems can be solved in three ways:
- Increase of computing power (e.g. parallel processing, special hardware)

- Write additional "high level" models for components, verify that the corresponding low level structure of behavioural elements fulfils the same logic function as the component ("stand alone mode") and simulate at "high level"
- Find a way to create a "high level" type model out of a structure of behavioural elements ("low level") automatically

The last approach was developed in Boeblingen and has been applied successfully for the functional simulation in previous projects. With our Design Language we are somewhere in between high level and low level simulation. Our chip is partitioned into 20 to 30 smaller structures called partitions. Inside a partition, the logic is described not at the gate level, but at the register transfer level. For this level our compiler generates one behavior type model out of a partition. (Partitions also serve as units in the physical design process, see "6.2.2 Partitioning and Floorplanning" on page 285). A behavioral high level model for the whole processor system is the beginning of the design phase. This was used to debug the first set of test instruction strings (test cases) and of some processor microcode. Later in the process the complete design was described in the Design Language which allowed us to debug the other test instruction strings (test cases) and microcode parts on the "partition level", avoiding the necessity to maintain and coordinate two design representations. Simulation execution time was still adequate (about 2 simulated clock cycles per second on an IBM 3081 processor).

Partition model interconnections were usually done automatically via equal names of signals on model boundaries. An additional program was produced that extracts the partition interconnect information on physical boundaries out of the physical design data (system "wiring" data) files and produce the interconnection for the simulation models, if the models were on that same boundaries. In this way these "wiring" data could be verified by our functional simulation. Alternatively system wiring data could be generated out of the simulation model interconnections, once they had been verified.

3.6.4 The Phases in Logic Design Verification

There are basically three phases in our design verification:

- Phase 1 is the time of preparation, where we provide the test instruction strings (BUSTER test cases) based on the design specifications and verify them on a high level model written for this purpose. The programs for model generation and production mode logic simulation are written and tested (the project dependency is added here). Phase 1 ends when the manual design process is complete.

- Phase 2 is the time of debugging the design. The first test instruction strings are simulated on the initial design model. At this point usually a large number of errors show up, either in the logic area, in the test instruction strings or in the "outside world" models.

- Phase 3 is the time when most of the errors in the design have been found with the existing test instruction strings (about 90%). All models and the test instruction strings are "clean". Now we run automatically randomly generated test programs on system level (equivalent to the S/370 architecture interface) in the hope to detect all remaining errors before we release the design to a silicon manufacturing line.

The requirements to the logic simulation vary from phase to phase. In phase 1 neither the model build process nor the simulation speed is a problem, since only a behavioral high level model is used which runs about 1000 times faster than the appropriate low level design representation, and the model build time is negligible. However it is very important to have a good interactive debugging feature for the test instruction coders available. This is also important in phase 2, but here its major purpose is to help the logic designer, resulting in different requirements. In this phase, the model build time (the turn-around time from error detection through correction to re-simulation) is the gating productivity factor. Simulation performance starts to become more important: the fewer errors are found, the more test instruction strings must be simulated to either detect the next error or to prove the last change did not hurt other functions (regression). In phase 3 finally the crucial parameter is simulation performance. Thus, only a go / no-go statement (without any traces) is produced for each individual test program. In the case of an eventual error, the simulation of that test program is repeated with the full debugging environment enabled to detect the error. To improve the simulation performance for phases 2 and 3 we used two different approaches: The first one tries to separate between the instructions within a test string according to setup or result check and between functional instructions (see "3.6.6 The Testcase Execution Control Program" on page 189). The second approach was a brute force approach to use increased compute power on multiple large machines for overnight and weekend shifts. As a result of both, we were able to simulate about 500 million clock (machine) cycles of test program execution on our simulation model in phase 3.

3.6.5 What Drives the Simulation - Testcases

The stimuli to drive our simulation model (representing a processor) are sets of microcode or machine instructions that build "testcases". These testcases (BUSTERs) can run on the simulation model by being loaded into the corresponding models of microcode memory and main store. They run in the same way on the real hardware (see "7.3.3.1 Sub-Architectural Verification" on page 322). This creates the capability to logically compare the simulation model with the real hardware. The test cases are built such that they are "self contained". There is no need to cross-check signals or to inspect traces. The results obtained by the simulation of logic gates and memory elements to fulfil some instructions are inspected by other instructions that compare with predefined data.

Logic Design Verification

As already mentioned, we link several models together to build a "system" to be simulated. For a functional simulation, the verification of design units may take place within a "system" of partitions (see "3.6.3 Modelling Requirements" on page 186) linked together, because the stimuli that are applied to the whole system are the same for any unit under test and the other units automatically provide the correct stimuli for the specific interface to the specific unit. With this method there is no need for a "stand alone" verification with stimuli to be generated exclusively for the one unit.

The amount of BUSTER test cases generated in phase 1 is based on experiences with previous projects. They are also used as the main vehicle to bring-up and debug the hardware (see "7.3.3.1 Sub-Architectural Verification" on page 322). Although there is no measurement of test coverage, we usually reached a coverage of about 90% at the end of phase 2.

The testcases for phase 3 are randomly generated strings of machine instructions (at the S/370 architecture level) requiring the complete microcode to be loaded into the model, too. The results are also pre-generated for these strings, so that these testcases are also self contained and are able to run on the hardware as well. With these testcases we were able to identify another 8% of the design errors. The remaining ones were found only on actual silicon hardware after running the operating system. They included no "show stoppers" that would prevent further testing of the hardware.

Historically, some microprocessors have been delivered to end users with design errors that were detected only later on. IBM ensures the design error free execution of S/370 programs. The very extensive amount of logic design verification contributes significantly to the integrity of the design.

Special tests were also generated that would test out the processor facilities in case of a component malfunction like parity errors, protocol mismatches etc. These tests could only be run on the simulation model, consisting not only of instructions, but also of some simulation control commands.

3.6.6 The Testcase Execution Control Program

The control over the execution of testcases is done by a special monitor program written in the Simulation Control Language described in "3.7.4 Simulation Control" on page 198. Since the Simulation Control Language has access to all signals and memory elements in the simulation model representing the design, the monitor can easily look into the model to control the execution. An example is shown in Figure 128. Important are errors, testcase start and good end. These events are signaled within the testcase by special opcodes that have no influence on the design function (so called "nops"). These "nops" are part of the testcase code.

Besides facilities for the execution of instructions, the CPU chip has a special "Unit Support interface" (see "2.11.6 Support Interface" on page 141). It performs functions like initialisation after power on, or Initial Microprogram

Load. This Unit Support interface was also simulated and driven by special Simulation Control Language programs.

The simulation of instruction streams also provided a certain signal pattern on chip boundaries. These can be captured by another part of the Simulation Control Language program and can be applied to feed a hardware tester to check out individual chips of the Capitol chip set with functional pattern in a stand alone mode.

3.7 Logic Simulation

Wolfgang Rösner

3.7.1 Overview

The software simulation system described in "3.6 Logic Design Verification" on page 184 consists of code-generating compilers for several hardware description languages and a simulator for execution and control of logic simulation. Figure 122 shows the dataflow of the simulation system. The entry languages for the system are IBM's Boeblingen Behavioral Language [WEBR], the register-transfer level Design Language (see "3.2 Hardware Design Language" on page 148), and BDL/S. The latter is IBM's hardware specification language for the physical design system EDS.

Conceptually a hardware system is modelled for simulation as a network of interconnected blocks. The block is the basic building block of the structure to be simulated. It may describe a complete system, or an individual logic gate. A block is described by

- Input signal lines
- Output signal lines
- Internal storage elements
- The behaviour of the block, its transfer function. This is the reaction of the hardware contained in the block to changes on its input signal lines and/or changes to the contents of its internal storage elements.

The interconnected blocks communicate via their signal interfaces. This modelling concept is identical to the nets-of-agencies approach in [WEND].

The domains of the signals in the system model are binary values in general and multiple discrete values in special cases. The system is performance optimized to simulate digital hardware on several levels of abstraction: the system level (functional/behavioral level), the register transfer level and the macro level (gate level, see Figure 99).

For the simulator, the transfer function of each communicating block is specified by one or several pieces of sequential S/370 code. Such a piece of sequen-

Logic Simulation

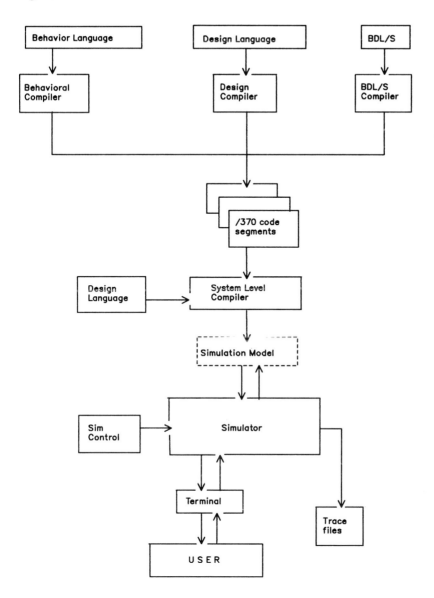

Figure 122. Simulation system structure

tial code is called "action"; a group of actions is characterized by the fact that they may be executed in parallel. Hardware designs usually have the characteristic that several pieces can work in parallel, and it is important, that a Behavioral Language description reflects this feature.

In our system the simulation is performed using an event-driven algorithm. The simulator contains an event-driven scheduler for block actions. Each time a blocks input changes, the corresponding block actions are scheduled to run.

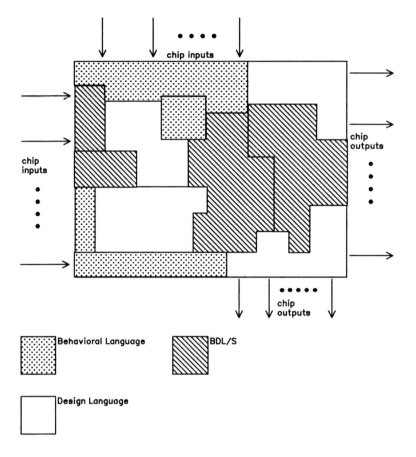

Figure 123. Mixed level simulation methodology

The algorithms are well-known and similar to those described in [BREU, RUSS].

At the top of Figure 122 three possible input languages to the simulation system are shown. To execute a description of some hardware in one of these languages, the corresponding code has to be compiled into a machine executable version. For each entry language of the simulation system a compiler is provided which generates S/370 executable machine code. Each block action is represented by one S/370-code segment.

There are compilers for two classes of hardware description languages. The first one is the Behavioral Language, resembling PASCAL but augmented with special hardware description and simulation features. This language reflects the block action structure of a hardware model directly on the surface. Thus it has a construct which allows the explicit specification of actions. The actions themselves consist of strictly procedural and sequential high level programming language code. The Behavioral Language Compiler transforms these actions into the equivalent machine coded simulator actions.

The second class of languages are the 2 structural languages: Design Language and BDL/S, i.e. register-transfer-level or gate-level languages. These languages simply describe a hardware system statically as a network of interconnected primitive blocks. The behavior of the blocks is either a function of Boolean logic, a storage or a simple value transfer function.

The compilers for these languages have to map the structural description to one or more block action programs.

A complete simulation model is built up by linking together the compiled hardware specifications written in different languages and thus originating from different levels of abstractions.

Figure 123 shows a single chip simulation model consisting of partitions. Each partition originally was described in one of the 3 languages. Inputs were compiled by the corresponding compilers. The mix of S/370 code segments originating from different source language descriptions was subsequently linked to generate a mixed level simulation model of the chip.

The linking is done by the System Level Compiler (see "3.2.4 The Macro Level" on page 154). As it happens, the Design language compiler (with a corresponding compile time option) is used for this purpose. On the basic hardware level the designer uses primitive built-in operators (e.g. Boolean functions) whereas on the system level (i.e. chip, card or board level) the building blocks are compiled blocks (Design Language or Behavioral or BDL/S) themselves.

Technically a Simulation model description gathers all the block action programs, all these segments of S/370 code which represent the behavior of the hardware model. The Simulator which also has knowledge of the statical system structure executes the block actions using its event driven scheduler.

The simulator consists of two conceptual layers. The basic layer is the scheduler. On top of the scheduler is a simulation control component which is an interpreter environment for the procedural "Simulation Control Language". This language is a high level programming language which gives the user interactive access to all model elements and it is also the user's interface to the scheduler. "Running" the simulation for a certain amount of model time is just one of the Simulation Control language statements. Simulation Control statements come either directly from the terminal of from procedure files (see also Figure 127).

The Mixed Level Design Verification System software runs under the IBM VM/SP operating system.

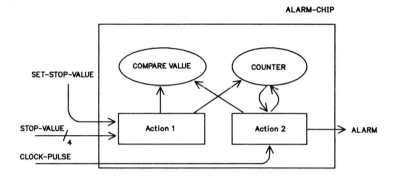

Figure 124. Alarm chip

3.7.2 Hardware Specification Languages

We restrict the following discussion the Behavioral Language. The Design Language has been covered in chapter 5.2. The gate-level language BDL/S contains a rich set of features for the physical design process. For details see [DONZ, MAIS].

The Behavioral Language is a procedural language with a PL/I- or PASCAL - like syntax [WEBR]. Its concurrency concept are socalled "actions". It includes hardware specific datatypes like BIT, BIT(..), ARRAY of BIT, and simulation specific elements like TRIGGER, WAIT, and AFTER. There is also a model time concept.

Figure 124 give an example of an "alarm chip" that counts CLOCK_PULSEs. Figure 125 shows the corresponding Behavioral Language example. The block is initialized by some outside hardware (or by the user via the Simulation Control Language) through the input signals STOP_VALUE (which is a binary coded number) and SET_STOP_VALUE (which is a synchronization signal). Normally the simulator schedules a block action when one of its inputs changes. "Input" means that the signal is a block input which is used in reading mode (i.e. in an expression) inside the action. The trigger-concept allows explicit specification of the signals that are used as triggers. This is used in both actions in Figure 125 (e.g. @SET_STOP_VALUE).

The explicit usage of simulator services is also shown in Figure 125. The CLOCK_PULSE ACTION issues a WAIT statement to suspend the execution of this action for 3 nanoseconds model time. The AFTER statement instructs the simulator to assign the value 1 to the signal ALARM with a delay of 1 nanosecond.

Logic Simulation

```
BLOCK ALARM_CHIP
INPUT
   STOP_VALUE (4),
   SET_STOP_VALUE,
   CLOCK_PULSE;
OUTPUT
   ALARM;
VAR
   COUNTER (4),
   COMPARE_VALUE (4);
/* ----------------------------- */
@ SET_STOP_VALUE ACTION
   COMPARE_VALUE := STOP_VALUE;
   COUNTER := 0;
END ACTION;
/* ----------------------------- */
@ CLOCK_PULSE ACTION
   IF COMPARE_VALUE ¬= 0
   THEN BEGIN
      COUNTER := COUNTER + 1;
      IF COUNTER = COMPARE_VALUE
      THEN BEGIN
         WAIT 3 NS;
         AFTER 1 NS ALARM := 1;
         COMPARE_VALUE := 0;
         END;
      END;
END ACTION;
/* ----------------------------- */
END BLOCK
```

Figure 125. Example for behavioral language

3.7.3 Compilation Techniques

The Behavioral Language Compiler is a regular language compiler. The main structural construct are the action declarations. These are compiled one-to-one to machine code block actions which are compatible with the simulator interface.

The Design Language compiler is more sophisticated with respect to code generation. Figure 126 shows the logic network for a design with three master- and slave- latches. To process such a design the parser of the compiler first generates an internal data structure which represents the network structure shown. Code generation is the task to map the network of primitive blocks (latches, combinatorial logic) to block action programs.

Two alternate solutions are conceivable. One is the compiled-code approach where everything is compiled into one large block action program. Here even a few signal changes in the model result in the re-execution of ALL the simulation code. Secondly it is possible to map each logic network block (e.g. AND gate) to one separate action. This totally event-driven method leads to a significant scheduling overhead.

The Design Language and BDL/S compilers solve the trade-off between parallelism and scheduling overhead. The network is partitioned into only a few parts whose interface signals have certain logical attributes according to LSSD design rules. As indicated before, these attributes are propagated throughout the combinatorial logic. The pragmatic approach here is to gather all master-latch input logic, all slave-latch input logic and all logic driving primary output each into one sequential block action program. Thus, one piece of hardware is mapped to 3 block actions taking advantage of inherent parallelism in hardware but avoiding too much scheduling overhead.

The result of the code generation process is shown also in Figure 126.

Besides the block action programs, the compiler supplies the simulator with the information how different blocks (designs) are interconnected. The simulator has thus the knowledge what action programs have to be scheduled when some input changes its value. E.g. if input 'a' in Figure 126 changes, the master-latch action and the output action are scheduled to run. Note also that the simulator provides service routines for the block action programs. E.g. the master- and slave-latch actions are able to call the simulator to schedule the slave-latch or output action resp. when one of the action-crossing signals changes. This again saves execution time since the simulator has not to evaluate, whether an action and which action has to be scheduled.

Logic Simulation

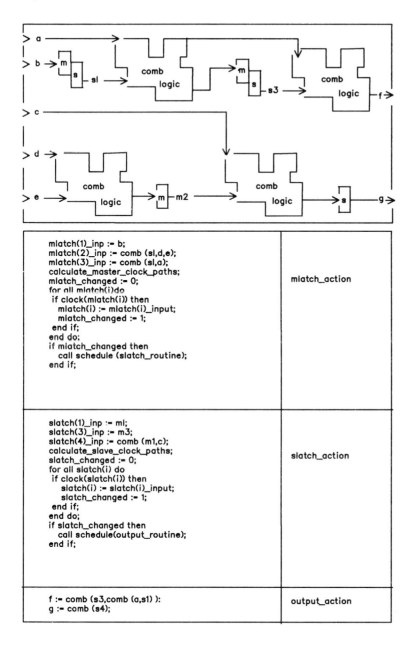

mlatch(1)_inp := b;
mlatch(2)_inp := comb (sl,d,e);
mlatch(3)_inp := comb (sl,a);
calculate_master_clock_paths;
mlatch_changed := 0;
for all mlatch(i)do
 if clock(mlatch(i)) then
  mlatch(i) := mlatch(i)_input;
  mlatch_changed := 1;
 end if;
end do;
if mlatch_changed then
  call schedule (slatch_routine);
end if;
	mlatch_action

slatch(1)_inp := ml;
slatch(3)_inp := m3;
slatch(4)_inp := comb (m1,c);
calculate_slave_clock_paths;
slatch_changed := 0;
for all slatch(i) do
 if clock(slatch(i)) then
  slatch(i) := slatch(i)_input;
  slatch_changed := 1;
 end if;
end do;
if slatch_changed then
  call schedule(output_routine);
end if;
	slatch_action

f := comb (s3,comb (a,s1)); g := comb (s4);	output_action

Figure 126. Action compilation

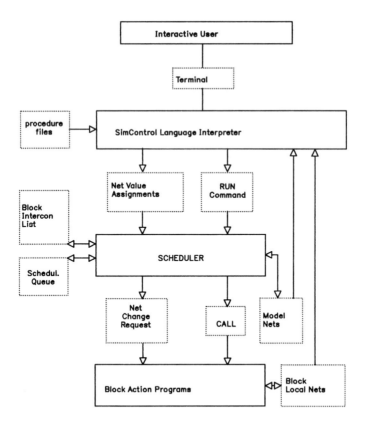

Figure 127. Structure of the simulator

3.7.4 Simulation Control

The structure of the simulator is shown in Figure 127. The user has access to the simulator via the Simulation Control Language. It is similar to PASCAL but has certain special features dedicated to simulation. Starting the simulator loads the model, which includes a symbol table that enables the control environment to refer to model elements symbolically. The model elements automatically belong to the scope of the Simulation Control Language. The user is able to assign values to inter-block nets and can display the current value of all nets in the model.

Figure 128 shows an example of a single Simulation Control procedure which stimulates the adder model shown in Figure 101 with all possible input patterns. From the simulation control level the user has access to all hardware elements represented on the model level. Signal values can be examined or changed. Running the simulation is controlled by one single statement of the Simulation Control language.

Simulation Control allows the user to define tracefiles (including their layout), breakpoints which can stop a running simulation, and also has a procedure

Logic Simulation

construct. One important statement is the RUN command which starts the scheduler.

The language has several constructs that opens user access to the underlying operating system. Especially, there is an interface to the VM/System Product Interpreter REXX [COWL, IBM4]. REXX is an interpreted language that issues commands to a subcommand environment. Default environment is the operating system's command language. The simulator is able to act as such a subcommand environment which makes Simulation Control Language open to all operating system services and commands. One of the features that can be used is an inter-user communication service which extends the simulation system to distributed simulation.

3.7.5 Distributed Simulation

Normally, simulation of a hardware model is done on 1 virtual machine (1 VM/SP userid). The hardware design language compilers and the simulator do not impose a principal limit on the size of the model.

However, there are reasons to apply more than one VM/SP userid to the simulation of one model:

1. Separate parts of a model that is to be simulated were designed with different logic design tools sets, i.e. different design languages for different simulators.

2. Simulation Control programs for separate, quite independent parts of a model are better kept on separate virtual machine to make debug and maintenance of both the Control programs and the model easier.

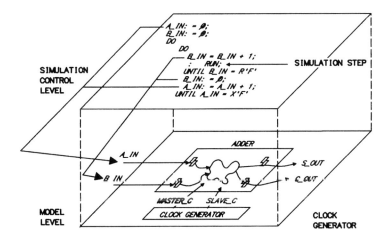

Figure 128. SimControl example

3. Simulation model plus the Simulation Control procedures which are required to run the model need more than the 16 MByte of virtual storage available under VM/SP.

Figure 129 shows co-simulation on two virtual machines.

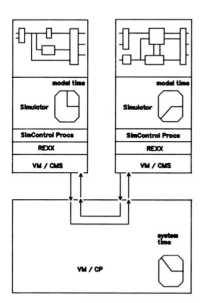

Figure 129. Concurrency in co-simulation

Part 4. Chip Technology

4.1 Chip Technology Overview

Gunther Koetzle

When defining the design concept of the Capitol chip set, three major decisions had to be made 4 years prior to the first product chips. They concerned the technology (CMOS or bipolar), the circuit library (small/simple or complex/macros), and the physical design approach (automated or more customized).

To get the right answers and to ensure an efficient development process for this complex task, a small, multi-disciplinary group of key professionals was established, responsible for the development of chip image, circuit library, I/O circuits, and Computer Assisted Design (CAD) tools (see Figure 130. The close cooperation within this group as well as with process technology, memory development, testing and other departments prevented sub-optimized solutions.

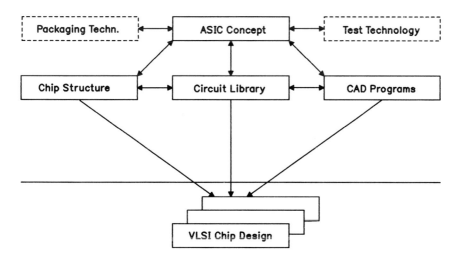

Figure 130. ASIC (Application Specific Integrated Circuit) development

A detailed analysis of the future product needs, the technological offerings and capabilities was performed prior to the definition of the design concept. Of particular importance was the analysis of the problems to be expected with the future VLSI technology and chip design. "4.2 Master Image Chip" on page 204 and "4.3 VLSI Book Library and Array Macros" on page 207 describe how these problems have been addressed by novel development/design approaches and tools.

4.1.1 Technology

The technology decision was easy for the technology group: CMOS. For the logic design team it was difficult to accept that a better performing system can be built with a "slower" CMOS technology, compared to the bipolar technologies used in earlier S/370 processors developed at the Böblingen Laboratory.

Figure 131 and Figure 132 explain the reason. The competitiveness of a product is not only - even not primarily - dependent on the speed of the basic circuits. The superior density of CMOS VLSI permits to reduce the "delay consuming" chip-chip and chip-module crossings in the critical paths by a factor of > 3. This permits to design the same machine cycle with CMOS circuits that may be 2.8 times slower than in a comparable design with bipolar circuits. Since the CMOS/bipolar basic circuit delay ratio is substantially less than 2.8, a better system performance results.

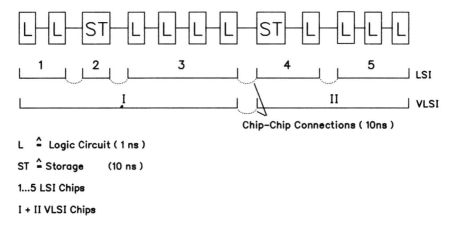

Figure 131. Critical path configuration

In addition, the cost of a CMOS based system is reduced due to lower packaging (less components, less power dissipation), and power supply cost (one voltage at 10% tolerance, lower wattage). Also, the reduction of solder pins, connectors and cables results in a better system reliability.

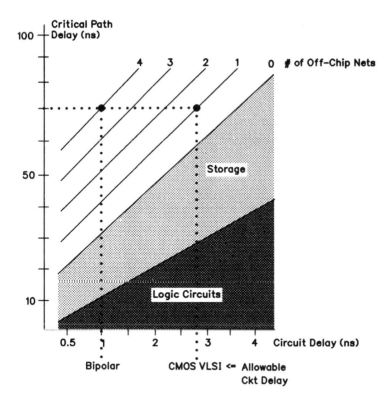

Figure 132. Critical path delay as a function of the level of integration at a given packaging technology

4.1.2 Circuit Library and Chip Image

The decreasing market life time in conjunction with rapidly increasing VLSI chip complexity requires a reduction in the development cycle. This forces a parallel development of process technology and circuits/chips, resulting in process groundrule changes during circuit design. A structured circuit design concept allows fast reaction to those inevitable process parameter changes. The structurization makes it feasible to generate a new circuit library based on a novel process within a few weeks. A small and highly structured circuit library - where all circuits are built up of a few common sub-circuit-elements as e.g. contacts, transistor shapes etc. - is mandatory for a VLSI product that is implemented with leading edge technology. Customization was consequently restricted to the dense array macros. No customized logic macro was designed or used.

Because VLSI logic chips are as a rule wiring limited (and not silicon limited), high chip density is achieved by "porous" circuits that provides a maximum of free wiring channels for the CAD programs. Due to the improved wirability there is no need to use the high-ohmic poly-silicon for wiring purposes, resulting in a chip performance improvement of 10%. The details are discussed in "4.3 VLSI Book Library and Array Macros" on page 207.

Since the available packaging technology (see section 4.3) has been optimised for bipolar chips with high DC and low AC currents, the high AC current requirements of the CMOS VLSI chips, and the simultaneously switching off-chip drivers limit the speed of logic and I/O circuits. To minimize the expected packaging delay, a novel driver with a digitally controlled signal slope allows a 2x better performance for the given packaging technology (see "4.4 A New I/O Driver Circuit" on page 211.

Optimization of the chip image for CAD requires regular structures. Aequidistant power busses of the same width on a fine grid are easiest for CAD. They result in a low effective power/ground inductance and thereby low noise in case of simultaneously switching circuits on the VLSI chip.

In addition, the regular structure of the chip image permits CAD programs to place custom designed array macros of arbitrary size and shape at any position on the VLSI chips. A detailed description of the chip image is given in "4.2 Master Image Chip." The corresponding physical design concept is described in Part 6.

4.2 Master Image Chip

Helmut Schettler

Structured (non-customized) chip designs use either the gate array or the master image approach. A gate array chip is divided into many cells of fixed size. Each cell contains the same logic gate (e. g. 3-way NAND gate). Multiple designs are implemented by a unique wiring of the basic cells with 2 (or more) layers of metal, using an identical underlaying silicon structure.

A master image chip personalizes the silicon in addition to the metal layers. It offers some of the advantages of Semicustom Gate Arrays and Custom Macro Design. The production turn around time is longer than that of gate arrays, but chip density and circuit performance approach those of the custom macro design. The Capitol chip set Master Image is very flexible and permits arbitrary size macros to be placed at any location on the chip.

The application of the appropriate CAD design tools allows a fully automated physical design at a small design effort. The Master Image is therefore very well suited for multiple part number designs. Also, transistors can be optimized for special circuit tasks. For example, the storage element of the Shift Register Latches needs special devices for performance optimization.

The chip image consists of a regular array of equally sized cell locations, initially without containing any active devices. Logic books (NAND gates, latches, etc.) implemented on the image consist of one or multiple cells (see Figure 133, showing the space requirement for 4 different books).

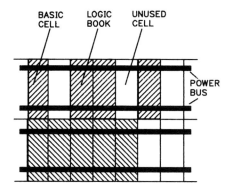

Figure 133. Master image chip cell arrangement

The cell locations are supported by a regular low ohmic and low inductive power distribution network, using three levels of metal. Logic books are connected by wires running within a regular grid of channels on metal1 and metal2. Power as well as logic wiring arrangement are optimized for VLSI chip density and CAD tools application.

The Capitol chip set technology is a 1.0 m CMOS N-Well structure with three layers of metal. Two layers are used for wiring while the third layer contains the power distribution and the I/O redistribution for the central area pad arrangement.

The chip pads are arranged on an area grid. All I/O redistribution signal lines are separated by power or ground busses for coupling noise reduction. The drivers/receivers are arranged at the periphery.

Some of the chip characteristics are:

Technology	CMOS	gate length N-device	1 μm
Chip Size	12.7 mmsq	gate length P-device	1 μm
Min. Feature	1 μm	max. no of transistors	800 000
Levels of Metal	3	Supply voltage	5.0 V
I/O Cells	232		
Wiring Channels	5010	typical delay 2way NAND	
Total Cell Count	98 800	C = 0.4 pf	0.8 ns

The design of the Image is tailored towards optimum wireability. The metal layers 1 and 2 are mainly reserved for global wiring. The chip design must minimize the area requirements for the power lines and the intra circuit connections. To minimize the circuit area itself is of second priority. It is the wireability of the chip that finally determines the chip's circuit or gate density.

The intra chip connections have to be located within the same wire channels for each circuit. This allows for continuous global wire channels all across the chip.

Metal 1 and metal 2 allocate the wiring area as follows:

	wiring area used on metal 1	metal 2
power distribution Vdd and GND	11 %	9 %
intra circuit wiring	14 %	0 %
global wiring channels	75 %	81 %

The Master Image is an array of cell locations without specifying gates or transistor locations. The active devices are positioned on a chip by the circuit placement program (see "6.2.4 Detailed Processing" on page 288). This method allows different circuits to make use of different transistor sizes and polysilicon connections. Especially the Shift Register Latches gain performance and wireability. The virtual cell grid is subdivided in a finer grid marking the vertical and horizontal wiring channels (see Plate 1). Actual wiring of a small part of the chip is shown in Plate 4a.

The periphery of the chip is occupied by 232 I/O cells. Each I/O cell may contain an INPUT and an OUTPUT BUFFER. The I/O circuits need special (large) transistors and they need protective devices against electrostatic discharge. The complexity of the I/O circuits (see Figure 137) requires many intra circuit connections. They cannot be laid out according to our wireability objectives. For this reason the I/O circuits are placed at the chip's periphery. The main area (84 % of the total chip area) is available for logic circuits and for macros (Data Local Store, arrays).

During personalization, the logic circuits and the larger macros are placed and wired by the automatic design system. A single logic circuit may take the area of one or multiple cells. Array macros are custom designed parts with a layout optimized for density and/or performance. The macros are either fully or partially blocked for global wiring. The power distribution grid on metal 1 and 2 of the Image is interrupted for macros. The macros contain their own power distribution and connect to the power grid of the image.

The image is suitable for circuit depopulation achieved by the placement strategy. During the placement procedure the chip areas showing a congestion of global interconnections are identified. They are depopulated by moving the circuits toward the peripheries of these areas. Empty cells in the center yield the required routing channels. Thus a uniform and optimal usage of the wiring channels is achieved. To allow this depopulation method, the chip area can theoretically accommodate approximately 30% more cells than can actually be used.

The power buses are distributed on special, somewhat wider channels, forming a low inductive grid on 1st and 2nd level metal. They are connected frequently to the 3rd level metal buses (see Plate 2). The supply current flows mainly through these wide distribution nets on the 3rd level metal rather than through 1st and 2nd metal. The effective ground-power loop inductance on the chip results in less than 150mV noise on GND and Vdd during current surges of 1 Amp/nS.

The 1st and 2nd metal power buses can be cut out anywhere to embed macros of arbitrary size. The macros may provide their own local power distribution, which is connected directly to the 3rd level metal.

Since the chip is mounted on a ceramic carrier using the Flip Chip Technique, the chip pads are located in a 4*4 mm area in the center of the chip. The small distance to the neutral point of the pads reduces the mechanical stress on the chip and pads during temperature changes.

4.3 VLSI Book Library and Array Macros

Otto Wagner

4.3.1 Cell Design

The Master Image is an array of cell locations without gates or transistors. This virtual grid is subdivided in a finer grid, marking the vertical and horizontal wiring channels. During personalization the logic gates (logic books) and the larger macros are placed and wired by the automatic design system. A single book may take the area of one or more cells.

Array macros are custom designed parts with an optimized layout.

Two design approaches for the cell are possible:

- The cell is split in an area containing the circuit with its internal connections. All automatically generated wiring is done in an extra wiring bay.
- The cell allows wires running on top of the devices. There is no wiring bay, and thus no separation between circuit area and global wiring area.

The second approach provides a maximum of chip area for wiring. Therefore this option was chosen for the Capitol chip set.

The basic cell size is 13.8 μ x 89.6 μ. At most 3 pairs of transistors fit within this cell. It contains 27 wiring channels on the 1st level and 3 wiring channels on the 2nd level metal. Two channels on the 1st level are blocked by power lines. (Channels numbered 8 and 20 in Plate 3a and Figure 136).

4.3.2 Circuit Library

All logic functions are implemented by a small library of about 20 basic books (I,A,O,AOI,OAI + Latches). Since VLSI chip circuit density is being limited by wiring capability - not by active silicon area - the logic books are designed for minimum blockage of wiring channels, (see "4.3.2 Circuit Library" on page 208). Depending on the complexity of the logic book, between 72% and 76% of all wiring channels crossing a cell are clear and free for logic wiring. The most complex book is a shift register latch. No customized macros are used for logic.

When a logic subfunction - equivalent to a complex standard cell macro - is implemented in simple basic logic books, a higher number of *short* connections have to be wired. The negative impact on wireability is overcompensated by a maximized number of clear and free wiring channels for logic wiring (6% loss versus $>20\%$ gain).

The summarized length of all connections < 0.5 mm is 1.75 m (or 6%) only of the 33 m total wire length used in the CPU chip (see Figure 134).

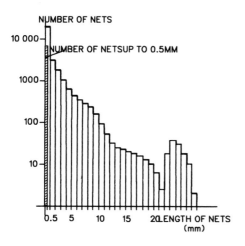

Figure 134. Distribution of wire netlength of a user representative 50k gates MIDS VLSI chip

The on-chip bus needs about the same wiring length.

All logic books are built from a few common simple structures (see Plate 3a). This constitutes an additional level of hardware hierarchy: the SUBCIRCUIT ELEMENTS. The resulting structured circuit design approach permits good flexibility and fast reaction to process groundrule changes, or new processes. It also reduces the required development effort.

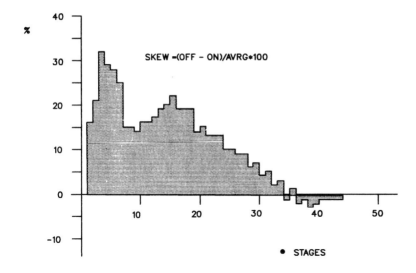

Figure 135. Skew compensation in a long logic path

Although no complex macros have been used for logic, it is possible to embed customized macros (logic or storage) of any size and shape. One example is the 72 Kbit cache array macro on the MMU chip. Macros are treated by the CAD programs like a large book.

4.3.3 Sub-Circuit Elements

The book layouts have to be aligned to the 1st and 2nd metal wiring, to match all groundrule restrictions imposed by any wiring and via combinations. This can be achieved by placing sub-circuit elements on the same grid as the wiring channels and connecting the pieces with standard wires of 1st and 2nd level metal, and of polysilicon. A logic function blocks 3 to 5 channels on the 1st level by its internal connections. For example, a 2way NAND uses 4 channels numbered 9,16,17,19 in Plate 3b, while a shift register latch needs 5 channels numbered 9,10,16,18,19 in Figure 136.

The method to use standard sub-circuit elements reduces significantly the design time. It allows late design changes. Thus, during the design cycle, groundrule changes and/or improvements can be carried out easily. In addition, the number of different elements can be kept small.

A single p type transistor and a single n type transistor are sufficient for all cell designs. Larger transistor devices can be made by paralleling basic devices. The driving capability of logic books is adjusted to load in steps. Books made with transistors of standard size show delays Ton and Toff, which are not equal. This is acceptable, because differences are compensated along a chain of logic books. This effect is demonstrated for a logic path of 43 stages. In Figure 135 the cumulated skew between a negative and a positive going transition propagating through the path is shown, which is in the range of 20% (relative OFF-ON/AVRG) after a few stages, but is almost zero after 30 stages. A pulse with a very short width shifted through such a path might disappear; this is prevented by the logical design of the Capitol chip set.

An exception concerning device size is made for the cross coupled inverters in the latch, which are made as a special element to achieve better performance with fewer devices.

A n and a p device connected to ground or supply voltage respectively insulates adjacent transistors. For safety against latch-up each book must contain an N-well contact and a substrate contact. The layout of the complete collection of sub-circuit elements is shown in Plate 3a.

A standard diffusion contact, a power contact to diffusion, a polysilicon contact, a 1st to 2nd metal via, a polysilicon connection combined with filling diffusion pieces completes the list of elements with which any desired logic book can be generated. In Plate 3b they are composed to generate a 3way Nand; in Figure 136 they generate a shift register latch with master and slave.

The circuit density on the chip is determined by the available wiring channels. There is no loss of silicon area by making elements not as small as the groundrule limits would permit. Therefore, elements are designed such, that the configurations of the sub-circuits match the wiring grid. Performance is limited by reliability, power and noise constraints. These considerations lead to a transistor size which is smaller than the cell height would allow.

4.3.4 Macro Design

Customized macros can be generated within short turn around times using the same method, but with a specific set of sub-circuit elements. This was done for the 48 x 36 fully static 3 port Data Local Store (DLS) array on the CPU chip. As the area of an array is defined by the dimensions of its silicon layers, the use of the channel aligned sub-circuit elements created a density loss of approx. 20 - 30%. The increase of the array area causes a loss in access time due to the larger parasitic capacitances.

However, a part of this loss is recovered as the macro can be adapted easily to the chip floorplan. The macro I/O 's and power lines match the pitches of the surrounding logic. Macro and logic books can abut without seam. The macro I/O 's can be rearranged optimally to the floorplan, e.g. to that side of the macro where they are best fitting to the interconnected logic. This reduces the usually large bay of interface wiring at the periphery of macros.

The increased area of macros designed by this method leaves clear feed through wiring channels, which can be used for connections across the array, e.g. from logic to the drivers at the chip edge. Generally this porosity reduces the length of wires and eases the wiring of global nets. The sum of macro area plus surrounding wiring area may be smaller with this porous macro approach.

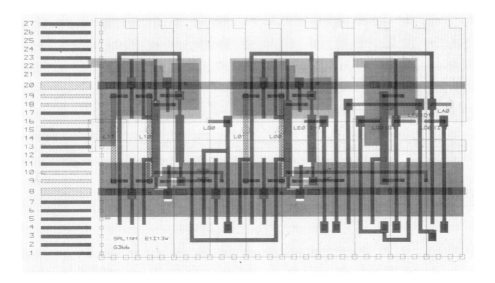

Figure 136. Layout of a shift register latch

4.4 A New I/O Driver Circuit

Thomas Ludwig

4.4.1 Problem Definition

The high level of integration in the Capitol chip set requires a large number of off-chip driver circuits for simultaneous communication between chips. Simultaneous switching relates to the number of driver circuits that switch at the same instant within a clock cycle.

CMOS circuits consume power only while switching from one logic state into the other one. Only during this part of the machine cycle a current flows from or into the power supply. This AC-current is mainly supplied by the on chip decoupling capacitance for fast switching operation. Especially off-chip driver circuits have the problem that the current return path of the output stage leads across the chip boundary and can not be compensated by on chip high frequency decoupling capacitors. The decoupling capacitor outside the chip is topologically the next source which can supply the required current. However, this capacitor can supply current only through the power and signal line inductances of the chip. Each current surge generates a voltage drop -Ldi/dt, which is proportional to the slope of the switching circuit output signal. The driver circuit environment is shown in Figure 137.

For a given maximum tolerable amount of noise due to output switching, either the number of fast simultaneous switching drivers has to be small, or the

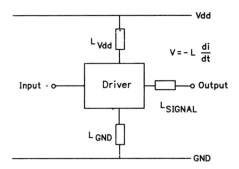

Figure 137. Driver circuit environment

drivers have to be slowed down. The circuit characteristics can spread over a wide range due to semiconducter process tolerances. The chips with optimum parameters in terms of performance (best case parameters, BC) show steep output slopes and generate the highest amount of switching noise, while chips with the poorest process parameters (WC) show flat driver output slopes. WC chips generate no switching noise problem, but their propagation delay is high. Variations of the supply voltage and the chip temperature add to the problem.

It is desirable to develop driver circuits which induce the smallest amount of noise into the power supply lines, in conjunction with an optimum propagation delay across all process and environmental conditions. This means that BC drivers are to be optimized in terms of noise voltage generation, and implies an increase in the rise and fall time of the driver output signals. If this optimization is applied, the off-chip delay for a WC driver will increase. The ratio of BC to WC driver delay is 1:3, which is unacceptable.

To solve this dilemma the Capitol chip set employs a new VLSI driver circuit family with special control inputs and a regulation logic that optimizes output driver signal slopes independent of chip process parameters, chip temperature, and supply voltage variations. The approximate worst case driver delay improvement by a factor of 2 is shown in Figure 138.

4.4.2 Driver Family

The new driver circuits are useable for CMOS Tristate driver (push-pull), and CMOS Open-Drain drivers. They use several control inputs which permit to adapt the slopes of the output signal. These control lines are connected to a control bus which guarantees similar rise and fall times for all driver circuits on the chip independent of process parameters.

The driver circuit block diagram is reproduced in Figure 139. Each driver consists of several building blocks: input logic, pre-amplifier, output stage, and ESD-protection. The input logic interconnects the driver data input with a gate signal. The pre-amplifier consists of several parallel stages. Two pre-

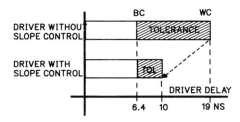

Figure 138. Off-chip driver delays

amplifier stages drive one output stage, and all output stages are interconnected at the output pin. The electrostatic discharge protection device (ESD) protects the I/O circuits against destruction during manual handling.

The control bus activates and/or deactivates additional pre-amplifier and output stages depending on the process parameters and environmental conditions. Activation of additional pre-amplifiers improves output signal rise and fall times, while deactivation slows them down. To guarantee the functionality, the pre-amplifier 0 is always active.

Best case (BC) chips require deactivation of one or several pre-amplifier and output stages. Worst case (WC) chips require activation of all stages. This assures identical output signal rise and fall times for chips with the best case process parameters and for those with worst case process parameters.

The book layout is based on the subcircuit elements described in "4.3.2 Circuit Library" on page 208. No other subcircuit elements are required to implement the logic and pre-amplifier stages. The layout of the drive stage and the protection circuit against electrostatic discharge uses special subcircuit elements. The subcircuit elements reduce the design time, and groundrule changes are easy to implement.

The I/O layout fits into the Master Image grid. All I/O cells are placed at the periphery of the chip. The cell height of a horizontally placed I/O cell is two times the logic cell height. The vertical placed I/O cells have a width of 10 logic cells. Each I/O cell contains an Input and Output Buffer. Depending on the used I/O cell type, a single Input or Output Buffer for unidirectional signals, or a combined Input/Output buffer for bidirectional signals is personalized by metal 1.

To improve the maintainability of the circuit library and reduce the design time, only two different driver types are implemented: A tristate-driver for normal usage of all non overlapping bidirectional system signals, and a open-drain-driver for special asynchronous overlapping control signals.

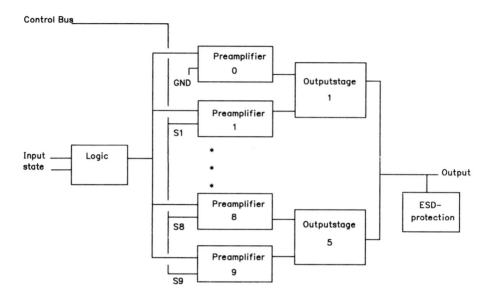

Figure 139. Driver circuit block diagram

4.4.3. Dynamic Control

On each chip is a single Digital Slope Control (DISCO) circuit that controls each output driver circuit via multiple control lines. A block diagram of this circuit is shown in Figure 140. The regulation is based on the comparison of the frequency of an oscillating loop with the system clock frequency. The oscillating loop consists of driver and receiver circuits. The driver is connected to a chip external capacitor equivalent to the typical driver load.

The oscillating signal generates an asynchronous clock that triggers a pulse generator, which defines a measurement period. The machine cycle counter accumulates the machine cycles occurring during the measurement period. A reference value is contained in the reference register, that is loaded during power-on initialisation.

The value loaded depends on the machine cycle time. Both values, the reference value and the machine cycle count result will be compared in the comparator. Three compare result bits(less, equal, greater) change the shift direction of the 9-bit-left/right control bus shift register. The nine outputs of the control bus shift register are directly controlling all off-chip drivers by activating or deactivating the driver pre-amplifiers as described. The nine control outputs also influence the oscillating loop and change its frequency depending on the number of active control lines. With this control of the oscillator the regulation loop of the DISCO is closed. Any parameter (voltage, temperature) changing the oscillator frequency during operation of the chip will lead to a change in the control bus shift register. Each chip contains its own DISCO circuit, running at the same regulated frequency. This frequency is independent

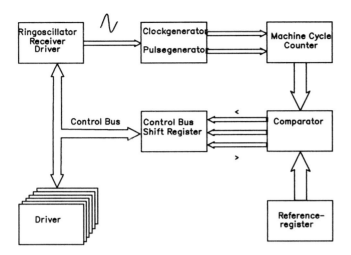

Figure 140. DISCO circuit block diagram

of process parameters and environmental conditions like supply voltage and temperature.

The DISCO circuit allows to fully utilize worst case characteristics without generating excess noise under best case conditions. To stay within acceptable noise limits, the driver must not exceed 25 mA/ns of current variation under best case conditions. The corresponding BC delay is 6.8 ns. Under worst case conditions the net propagation delay is limited to 16.7 ns at a 50 pF net load. A WC net delay of 7.8 ns with WC conditions is achievable, but results in a corresponding di/dt under best case conditions of 170 mA/ns.

With DISCO control the performance spread is 6.8 to 7.8 ns (instead of 7.8 to 16.7 ns), and the di/dt-spread is 23 to 25 mA/ns (instead of 25 to 170 mA/ns) over the whole range of process parameters, voltage variation and temperature. The delay-control driver circuit is estimated to have a 50% improvement in off-chip network delay compared to an uncontrolled driver circuit.

4.5 Embedded Array Macros

Klaus Helwig, Heinrich Lindner

The data flow of the MMU-chip as described in "2.4.3 MMU Chip Data Flow" on page 56 requires four array macros (arrays): Cache (high performance buffer memory), TLB (Translation-Lookaside Buffer for dynamic address translation), CD (Cache Directory), and Key Store. These four array macros are integrated together with the MMU control logic on a single chip (see Figure 141). Plate 8 shows a part of an MMU wafer.

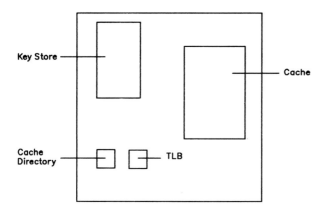

Figure 141. MMU chip with embedded memory macros

		CACHE	CACHE DIRECTORY, DLAT	KEY STORE
Organisation		4x256x72	4x32x17	4Kx8
Signals				
Data In	(DI)	72	68	8
Data Out True	(DOT)	72	68	8
Data Out Compl.	(DOC)	0	68	8
Array Select	(AS)	1	1	1
Addresses	(A)	8	5	12
Read/Write	(RW)	8	4	1
Data Gate	(DG)	4	0	0
Max. Power (mW)		700	90	170
Cycle Time (ns)		60	60	60
Redundancy		2 rows	–	1 row

Figure 142. Array configurations

Embedded Array Macros

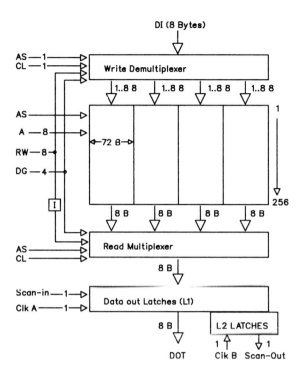

Figure 143. Cache

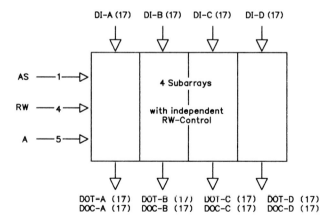

Figure 144. Cache directory/TLB

Besides density and performance, macro incorporation into the physical design system, easy testing, and a repair capability for manufacturing by redundant storage sites were major development objectives. The combination of above goals demanded individually tailored designs for each array; common building blocks (e.g. sense amplifiers), however, were shared between designs wherever possible.

4.5.1 Array Configurations

Figure 142 is a summary of array configurations as well as terminal signals.

The Cache in Figure 143 consists of four compartments (8 Bytes wide). The access (both read and write) is controlled by four data gate signals (DG). These are the four "late select" lines described in "2.4.4.3 Cache Array" on page 70. In any compartment, read/write of the 8 Bytes is determined independently by 8 read/write (R/W) signals. Row addressing is accomplished for the four compartments in common by the 8 address bits (A). The Cache-array is clocked by two timing pulses: Array Select (AS), starting row addressing, and Cache Latch (CL), setting the DG-valid time window see (Figure 146 for a timing diagram). AS, CL, RW, and DG are ANDed in the write-multiplexer or read-multiplexer to trigger column writing or reading, respectively. The data out latches L1 store retrieved data in between read operations. Clock A, clock B, L2 latches, Scan-in, and Scan-out are needed to allow the LSSD-test concept as described in "5.1.5 LSSD (Level Sensitive Scan Design)" on page 248.

The dual timing pulses, AS and CL see Figure 146 are required to allow late selection of the compartment by DG, once DG becomes valid after preceding TLB and Cache Directory accesses.

The Cache Directory (see Figure 144) has four compartments corresponding to the Cache compartments - with independent read/write control. Only row addressing by five address bits (A), common for all compartments, is required. Both data-out polarities are provided.

The functional requirements for the Translation-Lookaside Buffer (TLB, see "2.4.4.1 Translation-Lookaside Buffer (TLB)" on page 64) and Cache Directory turned out to be close enough that the same physical macro could be used for both; in the case of the TLB, read/write controls of two adjacent compartments are tied together. The penalty of four unused bits per row was accepted in favour of saving one macro design.

The Keystore macro (see Figure 145) is configured as 8 blocks of $4k \times 1$ bits, addressed in parallel with 12 address bits (A), thus, accessing the 7 data bits plus 1 check bit associated with each 4 KByte Block of real main storage. See "2.4.4.4 Keystore" on page 73. The output register follows the same principles as those of the Cache.

Embedded Array Macros

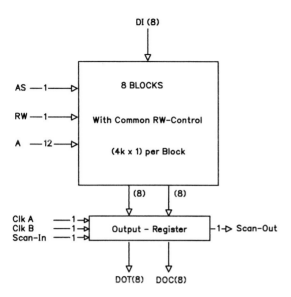

Figure 145. Key store

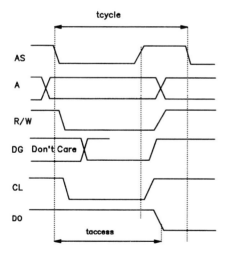

Figure 146. Cache timing diagram

4.5.2 Storage Cell and Circuit Design

The standard CMOS 6-device cell (see Figure 147 and Plate 4b) was chosen which offers a very low standby current, low soft error rate (critical charge 500 fC), as well as stability against electrical noise. The channel length of cell devices is 0.9 μm (actual). The cell size using minimum technology groundrules is 329 μmsq.

Besides the cell, other typical circuits employed in the three macros are:

- Sense amplifier (Figure 148)

  The Keystore Macro has a sense system which differs from that of the other Macros. The coupler devices P1 and P2 which are gated by the data lines DL and DLN isolate the sense nodes S and SN from the heavy data line capacitances. During standby, DL and DLN are in the high state, turning P1 and P2 off. Nodes S and SN are precharged to a low level by the reset clock RA, while the sense clock SA is high. When data is accessed, RA goes low. After sufficient signal has been developed across the internal nodes, SA goes low to set the sense amplifier. The signal is developed as soon as either level DL or DLN drops below the voltage Vdd minus VT. Therefore, the devices P1 and P2 can be designed large enough to speed up the pre-amplification. As shown in Figure 148 the sense amplifier can also be shared by two pairs of data lines via two pairs of coupler devices (P1,P2 and P3,P4). If two sub-arrays are sharing the same sense amplifiers, the bit line capacitance can be halved thus providing a faster sensing and a denser layout.

- Word Decoders

  Because of its small size, the TLB Macro employs the NOR decoder for the selection of the word lines. The NOR decoder is generally faster for small storage arrays with few addresses. Figure 149 shows a circuit diagram of the NOR decoder as it is used in the TLB Macro. Initially, the node WDD is restored to the high state, when the signal RWD is down. After RWD has been switched up, the five word addresses (true or complement) can be applied to the decoder. Only in one of the 32 decoders all 5 addresses are down: In this decoder, the node WDD stays high. After the signal EWD (Enable Word Decoder) has been switched up, the Word Line (WL) of the selected decoder is switched up, too.

  The Tree decoder used in the Keystore and in the Cache Macro offers low address input capacitances, small address buffers and a more favourable layout for larger arrays. Figure 150 shows a tree decoder as it is used for the Keystore Macro. The seven word addresses (true or complement) are applied to the X-tree and to the Y-tree resulting in 16 WX signals and 8 WY signals. These signals are combined to drive 128 word line drivers.

- Redundancy Steering

    Redundancy is activated by fusible links located at the chip periphery. The redundancy scheme and the circuit shown in Figure 151 provide an enhanced yield memory with no loss of access time.

    Figure 151 shows in particular a new address true-complement switch (T-C) which provides the memory with a distinctive feature so that the address inputs during the normal operation of the memory are not loaded by additional gate capacitances. There is no extra fuse, and thus no increase in DC power and time resulting from fuse blowing. If the chip is an all-good one, no redundant word or bit line WL or BL is required. Hence, there is no need to blow a fuse. Node C remains high to keep signal DSEL from staying low when signal ASW is activated. Transistors T17 and T18 are kept off by nodes E and F, so that address input A1C is isolated from the node A1TC. Similarly, transistors T15 and T16 are off, isolating input A1T from the thin-oxide capacitance, because nodes G and H are kept low and high, respectively, by nodes D and C and transistors T22, T23, T24 and T25. Thus, there are no yield losses attributable to downloading the address inputs. If a redundant word or bit line is needed, a fuse will be blown. Assuming the main fuse for nodes D and C is blown, then node C will be grounded, while node D is high. Transistor T1A is disabled, so that signal DSEL will remain low when the bad address is matched. For every address associated with the redundant word or bit line, fuse R1 will either be blown or be left intact, depending on whether the true or complement address input is selected. If the true address input A1T is selected, the fuse remains intact. The complementary input A1C remains isolated, while the signal of input A1T is transferred to gate A1TC by transistors T15 and T16. If input A1C is selected, fuse R1 is blown. The signal of input A1C can then be transferred to gate A1TC by transistors T17 and T18, as nodes F and E are high and low, respectively, at that stage. Node G is grounded and node H is precharged to high by nodes F and E. As a result, input A1T is isolated from gate A1TC.

    In brief, the address inputs are loaded by gate capacitances only if one or more redundant word or bit lines are activated.

Design objectives are met at 3-sigma statistical worst case for the process parameter spread, 10 to 85 degrees temperature range, and ±10 % supply voltage variations.

4.5.3 Array Integration

The design system described in chapters 5 and 6 imposes a set of procedures on the process of array integration. The array macros, though "manually" designed, have to be converted into objects that the design can automatically treat and process like any other element of the surrounding logic. In other words: the arrays, regardless of their large size, have to become equal "members" of the circuit library like basic circuits such as NAND and NOR.

This equal treatment of arrays at the various stages of the design system requires the following:

- Logic simulation: logic models for the arrays were generated in the behavioural language (see "3.7.1 Overview" on page 190). A considerable effort was spent to describe subtle device features realistically, e.g. warning messages in case of timing violations. Total chip and total system simulation was performed including the array models.

- Array/logic connections: signal assignments of array terminals were coded to allow subsequent automatic wiring.

- Placement and wiring (physical design): the layout at the array periphery needed several modifications for array integration: (see Figure 152).

    the array is surrounded by wide metal lines connected to the two power supply voltage levels; this acts as a noise barrier between array and logic;

    the signal lines to and from the array are connected to the surrounding logic at standard interface dots located on global wiring channels of the chip image;

    the area blocked for chip wiring is defined by blockage rectangles. Like any other device of the standard cell library, the arrays occupy an integer multiple of the image cell.

4.5.4 Testing of Embedded Arrays

By experience, testing of embedded arrays in manufacturing requires direct accessibility of the arrays through the chip terminals; in other words, the embedded arrays are to be testable like stand-alone memory devices. This was accomplished by multiplexing all array terminals to chip terminals under control of a special TEST-signal. If the signal is applied, the logic of the chip is conditioned such that certain chip terminals correspond 1:1 to array terminals. This mode is also used to determine defective storage sites repairable by redundancy.

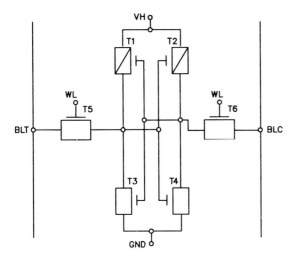

Figure 147. CMOS 6 device cell

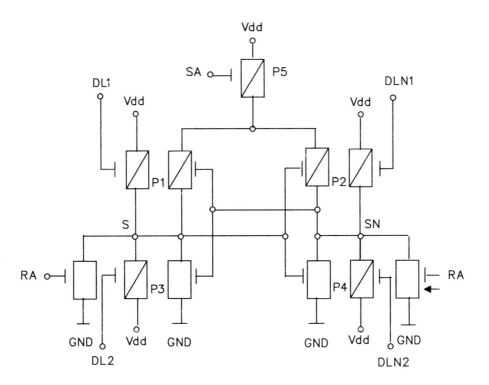

Figure 148. Keystore sense amplifier

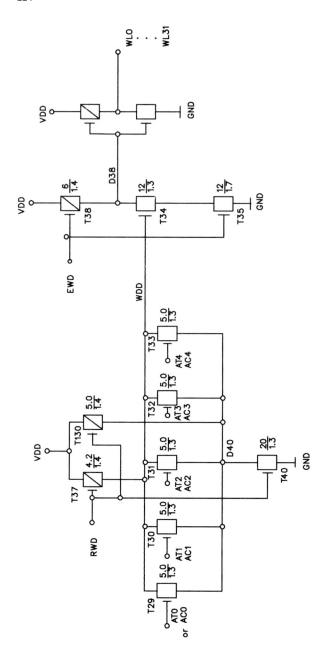

Figure 149. NOR decoder (for TLB)

Embedded Array Macros

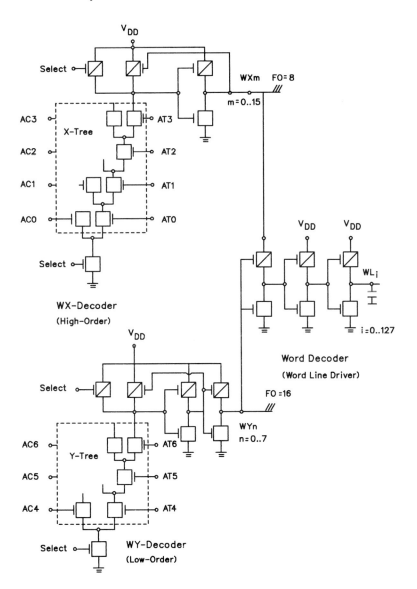

Figure 150. Tree decoder

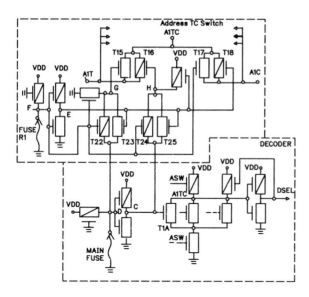

Figure 151. Redundancy steering

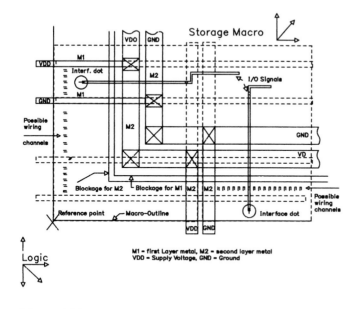

Figure 152. Array integration in logic. Special layout structures at array periphery (explained in text).

4.6 Packaging

Rainer Stahl

4.6.1 Overview

Compared to the fast change in VLSI technology, packaging parameters have only moderately changed.

	1979	1988
Chip Process Technology	bipolar	CMOS
Usable Cells Per Chip	700	60.000
Signal Inputs/Outputs Per Chip	96	164
Total Connections Off Chip (C-4)	120	439
Power Dissipation Per Chip (W)	1	1
Chip Size (mmsqu.)	20	160
Off-Chip Driver Switching Time (ns)	1	5

VLSI allows to integrate large functional units on single chips. The majority of signals leaving the chip belongs to address resp. data busses. The number of control signals is relatively low. The relationship of module solder interconnections vs. wired circuits on a chip has seen a 20-fold decrease in a decade. This results in a significant overall reliability improvement.

In the IBM ES/9370-30 and 50 realization the complete Capitol chip set processor is a functional and field replaceable unit packaged on a 17.7 cm x 22.6 cm glass-epoxy "Processor Card" of 0.12 cm thickness. 6 CMOS chips and 1 bipolar chip are joined to 28- and 36-mm-square "Metallized Ceramic" carriers. The remaining components are mainly standard dual-in-line packages as well as discrete resistors and capacitors. The organic card has a total of 516 connector tabs at the top and the bottom (Figure 153)

4.6.2 First Level Packaging

In the Capitol chip set the C-4 technology (controlled collapse chip connection, flip chip) is used to join each chip to a ceramic module. It allows to place the chip inputs/outputs basically everywhere on the chip. Wire bonding, in contrast, is restricted to the chip periphery. The C-4 technology utilizes PbSn solder balls, and a controlled collapse approach, to directly solder all chip connections to the module. With the small physical size and spacing of the PbSn balls very high numbers of connections can be realized. The CMOS chips are

Figure 153. Processor card

joined with up to 439 solder ball connections to metallized ceramic substrates. The CPU chip mounted on its substrate is shown in Plate 4b. Due to the low ohmic and low inductive nature of the C-4 joints superior power distribution is achieved.

Figure 154. Substrate with double layer metallurgy

The different coefficients of thermal expansion of the silicon chip and the ceramic carrier cause mechanical shear stress in the C-4 balls during temperature changes. The C-4 joints with the largest distance to the neutral point of the contact arrangement experience the highest stress. A thermal analysis of every module provides the details (rules) for the design of the C-4 pad arrangements. For the CPU, MMU and STC chip a "footprint" with a circular arrangement was designed (see Figure 154). The FPU, BBA and Clock chips have C-4 joints spread over the chips (see Figure 155). Calculations confirmed by thermal cycling experiments proved the superior reliability of the C-4 "footprints".

The line density in the device area requires minimum line widths and spacings of 0.050/0.063 mm on the ceramic carrier. The metallurgy is photolithographically defined thin film chromium/copper/chromium.

The metallized ceramic substrate process begins with a ceramic square on which the Cr/Cu/Cr films are deposited. Circuitry is defined by a photolithographical process using standard resist techniques and sequential etching methods.

Chromium is used for adhesion to the ceramic and as a solder flow barrier, copper for conduction. The circuitry is point-to-point wiring that connects semiconductor device pads (C-4) with input/output pin connections, which are inserted into the second level package. Figure 156 shows a detail of the substrate with double layer metallurgy insulated by a polyimide layer.

Figure 157 shows a detail of the substrate with single layer metallurgy. A detail of the double layer metallurgy is shown in Figure 158. A pin connection is visible in the lower left corner. In the upper half of the photograph there is first level metal. The horizontal line in the center of the photograph represents the step in the polyimide layer at the edge of the first level metal. The second level conductor crossing the photograph is following the topography of the polyimide layer. Prior to proceeding to the pinning operation, the personalized substrate is electrically tested for shorts and opens. Thus, defective substrates can be identified and rejected prior to final substrate processing.

The geometry of the device interconnections prevents visual inspection of 95% of the completed solder (C-4) joints. Therefore, careful control of the flux and solder reflow processes is required to secure perfect joining.

The final assembly process encompasses a polyimide coating to the substrate top surface, placement of the aluminum cap and an epoxy back seal that is dispensed on the underside as a liquid and cured to provide a "quasi-hermetic" seal. After marking and testing, the completed module is ready for attachment to the second level package.

Figure 155. Substrate with single layer metallurgy

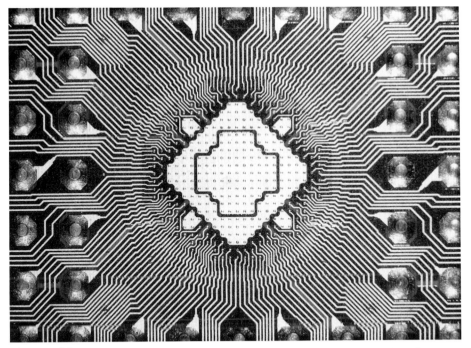

Figure 156. Double layer metallurgy fanout structure with chip connection dots

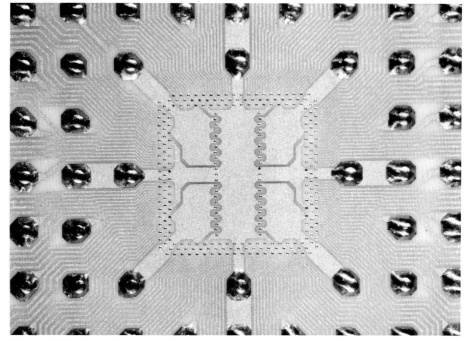

Figure 157. Single layer metallurgy fanout structure with chip connection dots

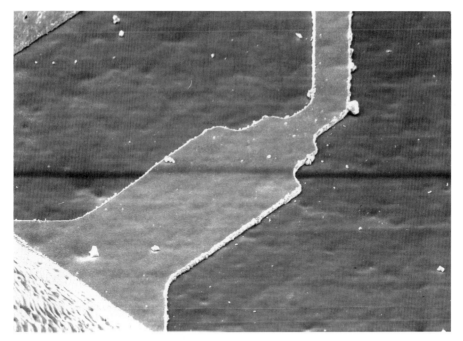

Figure 158. Detail of double layer structure

4.6.3 Electrical Considerations

CMOS circuits draw current only while switching from one logic state to the other. In contrast to bipolar circuits that create a relatively constant current flow, the CMOS circuits cause very steep current slopes.

The high frequency current demand during on-chip switching is mainly satisfied by the inherent on-chip decoupling capacitances. Recharging occurs through the carrier at lower frequency. The current for the off-chip driver circuits flows through the ceramic carrier to the decoupling capacitors on the organic card. The return current flows such that the inductive loop is minimized.

The carrier with double layer metallization shows very good electrical performance due to the microstrip implementation of the signal interconnections and the area layout of the power/ground supply.

4.6.4 Second Level Package

In the ES/9370 realization, the processor is packaged on a pin-through-hole type card with 4 signal and 3 voltage/ground planes.

The microprocessor chips are arranged in a fashion that guarantees excellent length (i.e. electrical load) control of the interconnections from the clock chip to the surrounding chips. In addition, the busses leaving the card are kept short by routing them from the chip to the nearest tabs of the card connectors.

The CMOS chips dissipate less than 1.5 W each. An airflow rate of 3 m/s is sufficient to keep the junction temperature well below its specified upper limit of 85 deg. C.

Newly devised "zero-insertion-force" connectors in the motherboards accept the 258 tabs on either card side. The card slides into top and bottom connector rails. The connector is actuated with a lever. During actuation, the connector springs wipe a specified length over the gold plated card tabs. This movement guarantees a reliable connection with very low contact resistance.

The total packaging concept is designed for very high reliability by providing burn-in for all modules and by minimizing the number of solder interconnections.

Packaging

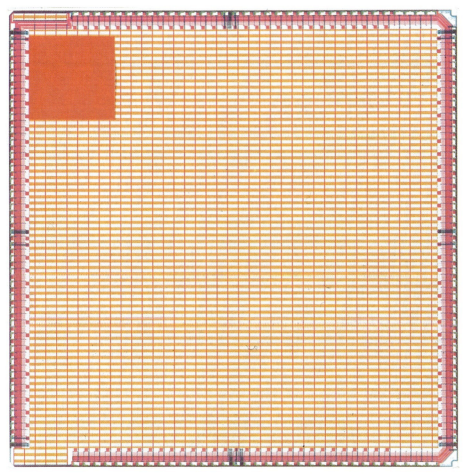

Plate 1 Master image with horizontal and vertical power lines

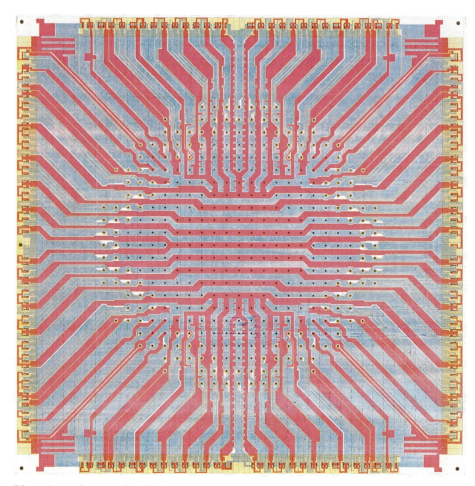

Plate 2 Power distribution on metal 3

Packaging

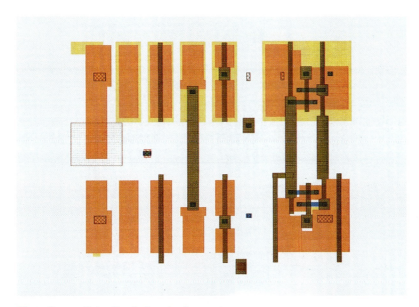

Plate 3a   Set of subcircuit elements

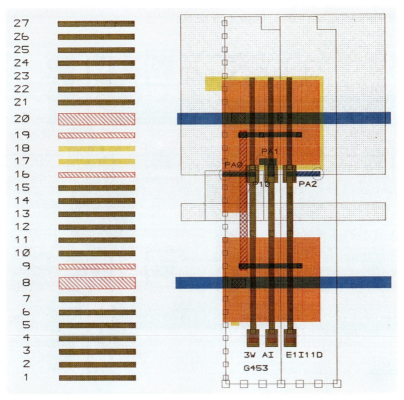

Plate 3b   Layout of a 3 way NAND

Plate 4a   First and second level wiring

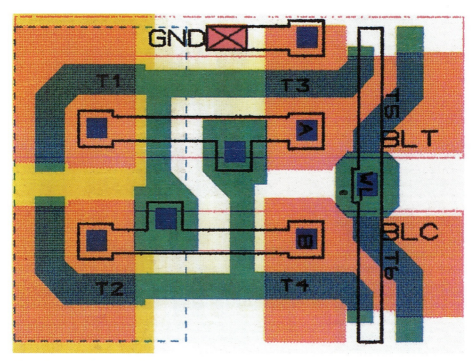

Plate 4b   6 device cell

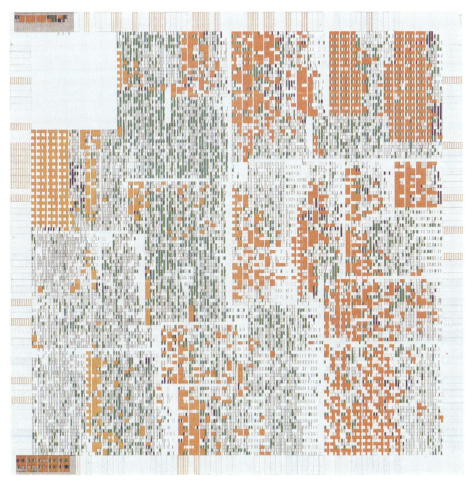

Plate 5 CPU chip placement

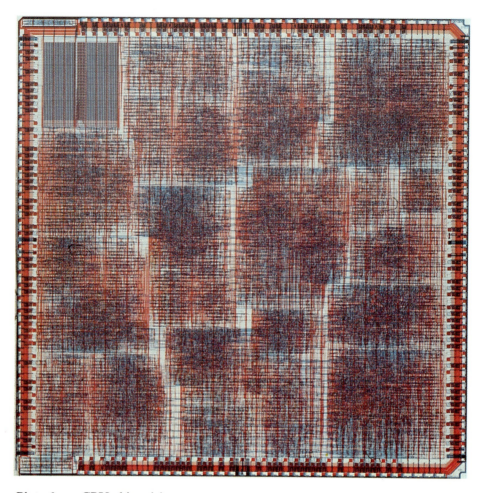

Plate 6   CPU chip wiring

Packaging 241

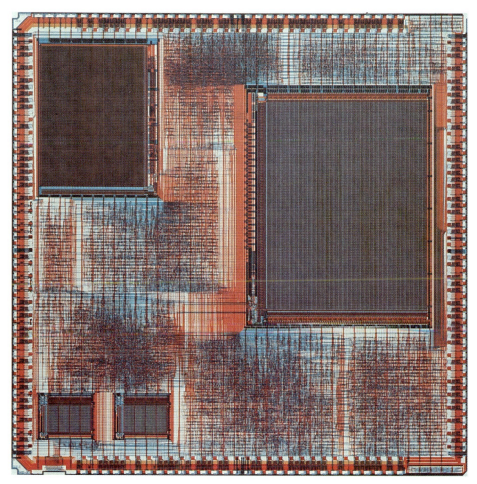

Plate 7   MMU chip wiring

Plate 8   MMU wafer

Part 5. Semiconductor Technology

5.1 Design for Testability

Michael Kessler

5.1.1 Overview

The high complexity of the Capitol chip set in terms of number of circuits requires special approaches to thoroughly test these parts. As a rule of thumb, the test effort increases with the square of the number of logic gates. The number of chip I/O's (input/output pins) has not increased proportionally with the circuit density. The new technology with low initial yields, new logic circuits, new embedded array macros together with expected new fail types further complicates the test and characterization task. Also, failing chips have to be diagnosed, and good chips evaluated and characterized in a short turnaround time.

A basic requirement for the development of complex VLSI circuits is a clear and high quality test method. It allows to introduce an aggressive new technology at an earlier point in time, pin point the cause of problems in the zero yield start-up time and later in the manufacturing time frame.

For these reasons, the Capitol chip set implements a strict adherence to "Design for Testability" Rules (DFT). Level Sensitive Scan Design (LSSD) is the main DFT feature, accompanied by a number of additional test features.

Prior to full function chips, several test chips were built. They were thoroughly characterized to support design optimization and process yield learning. An Advantest VLSI Tester was used to support the Chip/Module bring-up, the characterization, and the test specification development. Specific attention was devoted to the diagnostic capability through specific test features and diagnostic software support. Hardware diagnostic tools in the test area are described in section "5.2 Test and Characterization" on page 255.

5.1.2 Failure Types and Failure Models

A prerequisite for deciding on a test strategy requires reasonable assumptions as to expected failure types and their frequency. Underlying fault classes on silicon chips are discussed in section "5.4 Failure Analysis" on page 270.

Depending on the applied test method the failure types need to be represented in appropriate failure models. These models are an abstraction of reality. The level of abstraction is also dependent on the test method. Some of the failure models can be handled efficiently by computer programs. Commonly used failure models are:

- In the STUCK-AT-Model, In- and Outputs of logic gates are "Stuck"-AT "0" or "1" (Stuck-AT-Fault). This is the most important and widely used failure model. It was also the main fault model applied to Capitol chip set testing.
- The DYNAMIC-Model reflects the time needed from one gate to the next.
- The PATTERN SENSITIVE model is used to test memory elements. The various pattern sets test for a comprehensive set of memory fails.
- The FUNCTIONAL Model consists of the TRUTH- / STATE table of the circuit.
- The CMOS FAILURE Model is specific for CMOS Circuits. In the case of a CMOS-STUCK-OPEN fail, combinatorial logic can be transformed into sequential logic. The test pattern must take this effect into account.

The Stuck-AT-Model was used in most cases. The other fault models served to improve product quality. Some of the physical books were modified such that in the case of a CMOS failure, the fault effect appears like a simple Stuck-AT-Fault.

5.1.3 Structural Test

Test methods for integrated circuits fall into two classes: functional test, and structural test.

The functional test uses the truth table of the circuit and verifies the matching of all table entries with the hardware. Alternatively, the proper working of all functions / commands is verified. Unfortunately, in most cases the truth table is not complete (too large), nor is it feasible to test all commands with all their variations. Also, there is no good measure of quality available for these types of tests.

Because of this, the structural test was used for the Capitol chip set. It is based on the actual physical layout of the chip and reflects the circuit at the gate level. The Stuck-AT fault model is utilized. This model assumes ideal gates. Only the input and output net of a gate can have a logical '1' or '0' Stuck-AT fault. The Stuck-AT faults of all nets are listed in a failure dictionary. The number of modelled faults is not boundless as it is for functional testing, and is manageable for computer programs.

The structural test patterns are automatically generated as part of the physical design process (test data generation see "6.2 Hierarchical Physical Design" on page 283). As indicated in Figure 169, a special conversion program converts them into the "test vector" format required by the VLSI tester.

Some of the "Buster" test cases mentioned in "3.6.5 What Drives the Simulation - Testcases" on page 188 and "7.3.3.1 Sub-Architectural Verification" on page 320 were used as a supplement to the structural tests. These test cases are produced by the individual logic designers. They are used for simulation, chip testing with functional patterns ("5.2.1.4 Functional Pattern Test" on page 258), and system bring-up. Using the VLSI tester described in "5.2.1.5 Logic Test Equipment" on page 259, they can be applied at real system speed. Structural test patterns based on the Stuck-AT model can be applied on the same tester only at a limited speed.

The design for testability and the test methods described below are based on this structural test.

5.1.4 Design for Testability

As mentioned before, it is necessary to know the type and the quantitative distribution of the expected failures for each production step to choose the optimal test method. The cost to remove a failure from the product is increasing tenfold per packaging level e.g. chip, module, card, system.

The requirement to eliminate faults as soon as possible leads to a series of different tests per packaging level:

- Stuck-AT tests, complemented by AC-tests, for chips and modules
- In-Circuit tests on card level,
- Functional tests in the system.

To apply these different test methods effectively, the corresponding testability rules for VLSI chip designs have to be observed. The most important Design for Testability (DFT) rule tries to improve the controllability and / or the observability of internal nets. Through this, an assumed fault effect can be measured from peripheral chip pins.

The largest part of the computing resources for test data generation is spent for fault detection and fault simulation. Fault detection in sequential circuits is very compute-time and memory intensive and is not feasible for today's VLSI circuits. Functional tests are too limited to be used as the primary test approach. Fault detection in combinatorial networks is much easier than in sequential networks. All single Stuck-AT faults in non-redundant combinatorial networks can be detected. Required is a logic structure, which allows the network to appear as a combinational network within structural test data generation.

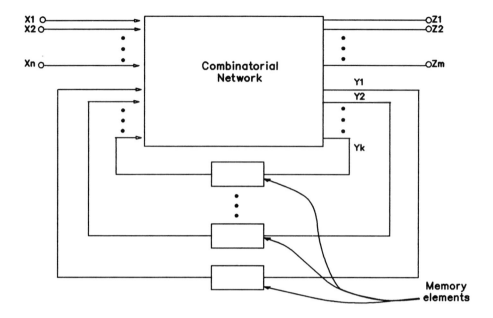

Figure 159. Huffman sequential network

5.1.5 LSSD (Level Sensitive Scan Design)

5.1.5.1 Overview

Level Sensitive Scan Design (LSSD) is one solution to the problem of test data generation in digital networks. The basic idea of LSSD is to make a sequential network look like a combinational network by logically cutting the feedback loops (contrary to cutting them physically which is costly in terms of precious I/O's). This logical dissection is performed by converting all memory elements in a Huffman Sequential Network (see Figure 159) into shift register latches (SRL's) and connecting all these SRL's into one or more shift registers (see Figure 160).

Now the network needs no homing sequence, since it can be put into any desired state by shifting the proper values into the latches. Also, if any faults propagates to a state output, the fault value does not have to circulate around the feedback loops as before. Rather it can be latched into the corresponding SRL and then shifted out for observation. Thus, the problem associated with fault detection in sequential networks has been resolved and the network appears like a combinational network as far as fault detection is concerned.

LSSD, by its structure solves yet another problem. Testing is considerably complicated by the presence of AC dependencies such as dependencies on transition time, switching speed and minimum circuit delay. However, in LSSD, all SRL's are assumed to be made up of DC latches. There are latches which react to

Design for Testability

DC logic levels and are not affected by transition times. Furthermore, clocks are recommended to be non-overlapping during system operation and are never overlapped during testing; hence the network becomes immune to fast paths since an early signal arriving at a latch has to wait for the next clock.

For example, in the two clock system shown in Figure 161, when clock C1 turns on, the signal through the fast path has to wait until clock C2 turns on. Since C1 and C2 are never on together, a complete loop through the fast path is never completed. There still exists some dependency on maximum circuit delay. The cycle time for a given network will specify what the maximum delay through it can be and coupled with a delay figure per logic block will specify how many logic levels a designer can put in the longest path between two latches. If the number of logic gates (blocks) in this paths exceeds the allowable limit, the network may not have sufficient time to perform certain functions within a machine cycle - hence the dependency on maximum circuit delay.

Another key feature of LSSD is the capability to completely control and observe all latches. They can be set from the outside to any desired state and similarly the state set into them can be observed. These two features are essential in making a sequential network appear like a combinational network. Hence, it must be possible to shift-in (or scan-in) and shift-out (scan-out) all latches in a design.

5.1.5.2 LSSD Rules and Partitioning

To allow the use of our Automatic LSSD Test Generation System, the VLSI logic design has to observe a number of LSSD rules. In addition, the LSSD rules need to be automatically checked. The rules are formulated such that a program is capable to check the design for rules compliance. One way to design a Shift Register Latch (SRL) is shown in Figure 162. Since L1 and L2 have separate clocks, the design is inherently level sensitive.
Memory elements outside of the LSSD shift register chains are only allowed as array macros. They need special attention.

A complete test of the array macros is done through the surrounding logic. In addition, the logic in front and after the array has to be tested as well. The latter is somewhat difficult, since the test pattern generation and the fault simulation program must propagate the fault effects through that array. Figure 163 shows the general interface between logic and arrays.

Several array integration rules guarantee and eases the test of the logic and the array. One of the most important rules is the 1 to 1 correspondence from the primary inputs (PI) or SRLs to the array and from the array to the primary outputs (PO) or SRLs. Figure 164 shows an example.

The compute time for test generation and fault simulation is proportional to the square of the number of logic gates in the network, see Figure 165. Splitting a large network into smaller sub-networks and computing the test patterns separately for each sub-network, results in a reduction of overall compute time. LSSD Partitioning is a way to subdivide a circuit into smaller sub-circuits. The

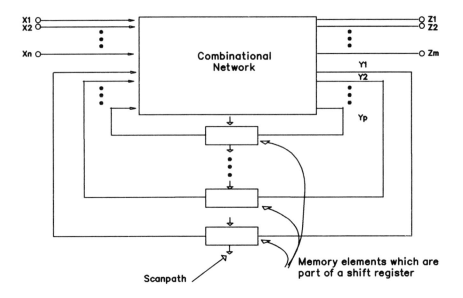

Figure 160. SRL's shift registers

algorithm starts at a primary output (PO) or SRL and traces back thru the logic until primary inputs (PIs) or SRLs are reached. This logic "cone of influence" is combined with other cones to form a reasonable size partition, see Figure 166.

5.1.6 Additional Test Features

5.1.6.1 Internal Tristate Driver

The CPU chip contains an internal bus, the Central Data Bus, see "2.2.2 Block Diagram Description" on page 20. This is a Tristate Driver (TSD) bus.

A DC-Stuck-At test cannot completely test internal TSD's, because the internal HZ-State (high impedance) cannot be measured directly. The HZ control logic remains intestable. In case of a stuck fault in that area, it is possible that two TSD's pull opposing levels on the same line. This can cause serious reliability problems, if those stuck faults go undetected into a product. To minimize the number of intestable faults, the HZ control logic must be fed to an observation point (SRL or PO).

The Stuck-AT test cannot test the HZ line of an internal TSD because of the incapability of modelling capacitances. A special test program can make use of the stored value in the capacitance of the TSD line, thus testing for the HZ stuck faults.

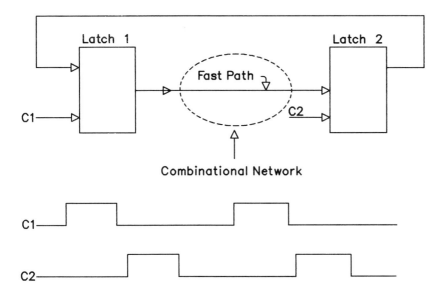

Figure 161. Non-overlapping clock structure in LSSD

5.1.6.2 Observation Points

The "zero yield" issue was addressed early in the development of the Capitol chip set. It relates to the following likely scenario: Let us assume a perfect logic and physical design following all the above mentioned testing rules. Assume one systematic design error on the first chips coming from the pilot line. How can one systematic failing transistor out of 800 000 be found? What happens, if beside the systematic fail there are also many random defects which change from one chip to the next?

To overcome this potential early Zero Yield phase various debug facilities were designed into the chips.

- All DC-Stuck-AT testing works only if the SRL registers are working properly. They are active test elements. In case they are not operational, additional debug features are needed. Therefore, special SRL observation points were implemented for diagnosis. Every 16 bit the Scan path was tapped and multiplexed to a PO (primary output).

- The chips are already fully logically testable after the second metal wiring. The third metal level is implemented only for power distribution and I/O redistribution and prohibits internal probing for diagnostic purposes.

- A special probe point book was included into the logic book set (see "4.2 Master Image Chip" on page 204) to be placed on critical chip nets. It allows easy access by internal probing of the circuit.

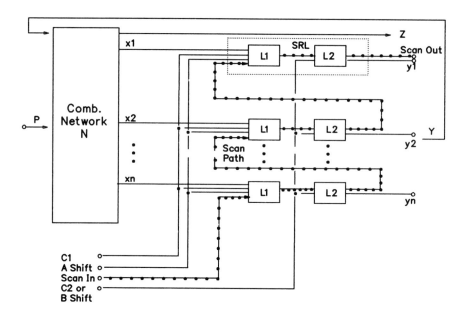

Figure 162. LSSD double latch design

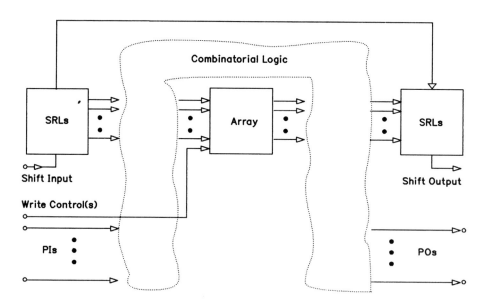

Figure 163. Interface between logic and arrays

Design for Testability

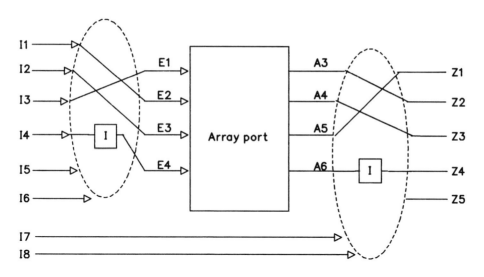

Figure 164. Input to and from array

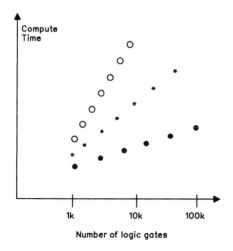

○ not structured design

* structured design

● structured design plus positioning

Figure 165. Compute time for test generation

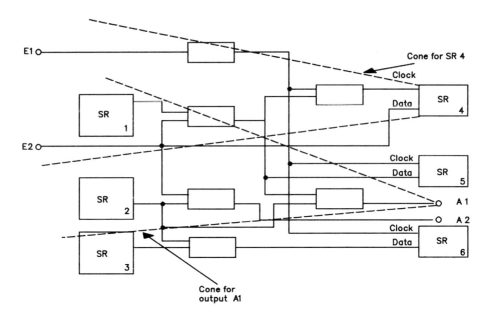

Figure 166. Logic cones form a size partition

5.1.6.3 Logic Circuit Layout Optimized for Defect Sensitivities

As mentioned in "4.4 A New I/O Driver Circuit" on page 211 the CMOS faults complicate testing, because a CMOS-STUCK-OPEN fault needs two successive test-patterns to be tested. Including those possible faults makes test generation and fault simulation much more expensive.

For this, the physical book layout was optimized. Most of the potential CMOS-STUCK-OPEN faults now appear like an normal Stuck-At fault, see "4.3 VLSI Book Library and Array Macros" on page 207.

Analysis and simulation demonstrated this to be an acceptable approach. It saved a significant amount of compute power.

5.1.7 Random Pattern Testing

The design for testability rules includes a section for pseudo Random Pattern Testing (RPT) in conjunction with signature analysis. RPT testing can be seen as one form of self-test. RPT patterns can be applied from very simple external test equipment. It might be included on the printed circuit board in the system. Figure 167 shows the basic layout of the LSSD chips. All scan-in and scan-outs were multiplexed with normal data pins, thus not requiring any extra I/O pins for scan-in and scan-out. Figure 168 shows a RPT tester environment.

These are some of the requirements for RPT testing:

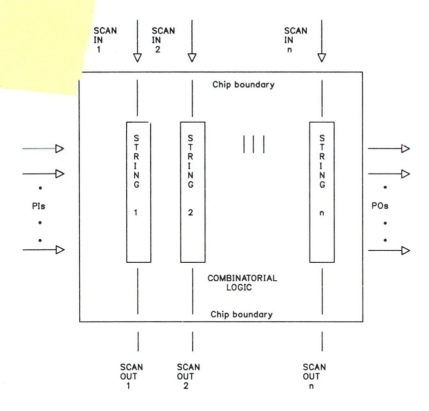

Figure 167. Basic layout of the LSSD chips

- Exclude the possibility that any unknown logic states in the network reach the signature compressor:
- Multiport Set-Reset latches require extra provisions to avoid undefined logic states for all random patterns.
- The Tristate Central Data Bus on the CPU chip features some additional controls to avoid "HZ" logic values to enter the signature compressor.
- The multiport Data-Local-Store on the CPU chip is designed to avoid unknown logic states (e.g. simultaneous writing and reading on one address resulting in undefined values).

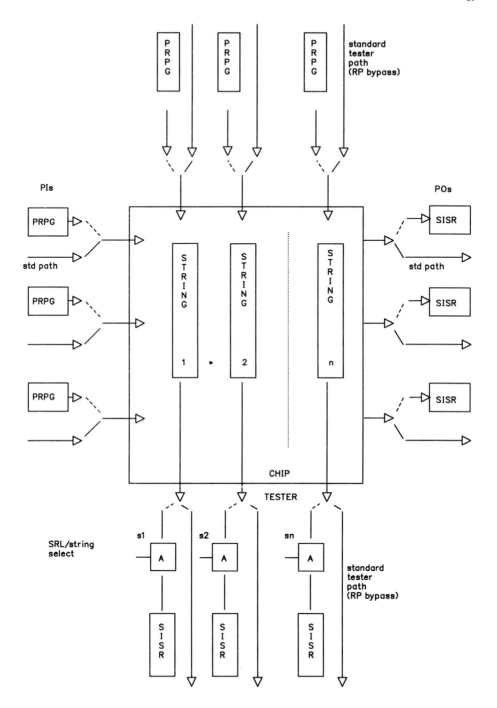

Figure 168. RPT tester environment

Design for Testability

5.1.8 Auto Diagnostic

Finding the failing net in a faulty VLSI circuit is a difficult testing task. To find process problems (mask errors, repeaters, contamination etc...), diagnostic tools (software plus hardware) are needed to pin point to the fail location.

In normal Stuck-AT fault testing, the fault simulator produces a failure dictionary for each detected fault. In many cases it is sufficient to use this failing pin / pattern information from the tester, to find the corresponding fault diagnostic in the dictionary and in turn the failing net.

Unfortunately, there are faults, which manifest themselves not as single Stuck-AT faults: CMOS faults, bridging faults, multiple Stuck faults, AC-faults, etc. If they are caught by the Stuck-AT test, there could be either no, or too many potential fault candidates.

In those cases an automatic software diagnostic tool was used. The tool takes not only the first failing pin / pattern information from the tester, but all the fail information till the end of the test. With that extra information the tool is capable in most cases to predict the failure location.

- In case of too many potential fault candidates (e.g. early in the test sequence) it algorithmically excludes potential fault candidates. The result is normally one or a few potential fail locations.

- In cases were no fault information was recorded in the failure dictionary, the possible problem was not a single Stuck-AT fault. Here, the tool permits to activate other options: it allows to model bridging, CMOS, AC and multiple Stuck-AT faults. The tool does also noise analysis.

During the development phase a one day turn around time from first time testing of a new lot, till identification of e.g. a mask repeater somewhere in the logic could be achieved. This capability, together with the physical internal probing was very effective in the Capitol chip set bring-up.

5.2 Test and Characterization

Joachim Riegler, Peter Hans Roth, Dieter Wendel

5.2.1 Wafer and Module Test

5.2.1.1 Test Overview

The Capitol chip set introduced a new semiconductor technology simultaneously with a new design. Once pilot production had started, a significant variety of test and characterization tasks were carried out on different levels of Capitol chip set engineering hardware. The test sequence of wafer and module

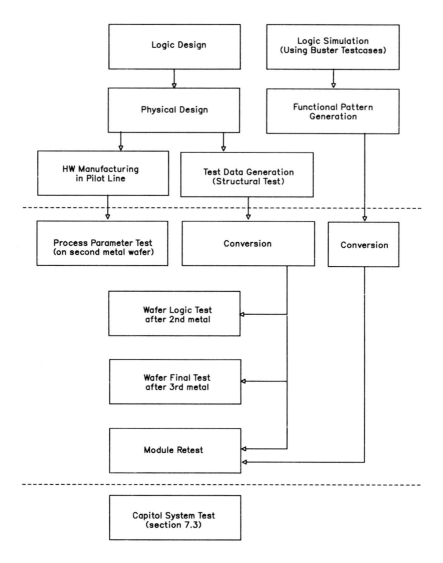

Figure 169. Capitol chip set test strategy

tests, their embedding in the flow of chip processing, and subsequent system tests are shown in Figure 169.

The complete test flow consists of three major parts:

- Specific process parameters are continuously checked and evaluated. This test is executed on wafers after completion of second layer metal processing.

- The logic function of the complete chip is verified. This test is mainly executed on wafers after third layer metal processing is complete.

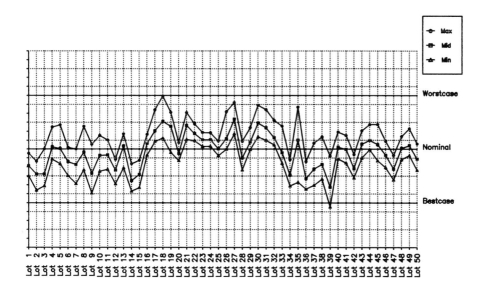

Figure 170. Channel length spread vs hardware lots

- The AC test utilizes functional patterns and verifies timing specifications on modules.

5.2.1.2. Process Parameter Test

During the development phase of the Capitol chip set, a wide variety of process parameters and yield detractors were closely monitored. For this, a special test site structure is attached to each chip, arranged on the wafer between the chips. Next to single transistors and storage elements, this test site holds technology oriented structures to verify the process parameters. For example, transistors are used to determine the actual channel length and width, threshold voltage and charge carrier mobility. Figure 170 shows the range of the channel length of n-channel transistors for multiple wafer lots over a certain period of time. A valuable data base can be created by continuous examination of the parameter test data and their collection during the complete development phase. Process parameter adjustments can be planned using this data base to optimise the processing points. In addition, parametric test results are used for performance verification and performance prediction (see also "5.2.2.4 Performance Verification" on page 262).

5.2.1.3 Logic Test on Wafer and Module

All parts of the Capitol chip set are designed according to Level Sensitive Scan Design rules (see "5.1.5 LSSD (Level Sensitive Scan Design)" on page 246). The flow of logic test pattern generation out of the design data is shown in Figure 169. The logic test requires test patterns (test vectors) that can be applied to the chip under test with a product specific voltage level, timing, and pin setup.

A typical test program may include the following components:

- During DC parametric test, the input and output circuits of the chip and their electrical behavior are analyzed. Leakage current measurements and three-state driver control are examples for parametric tests.

- During logic and array test, logic and array test vectors, obtained from the structural test patterns (see "5.1.3 Structural Test" on page 244), are executed in line with the LSSD concept. Different electrical conditions are set up for module and wafer level hardware to guarantee the application range.

- AC testing consists of propagation delay tests and functional pattern tests.

  In delay tests, the propagation time of pulses through typical and/or time critical logic paths is measured. These results serve as a performance indicator for the tested chip.

5.2.1.4 Functional Pattern Test

Section "3.6.5 What Drives the Simulation - Testcases" on page 188 explains, how the logical designers develop microinstruction sequences ("Buster testcases") to verify the correctness of their design through logic simulation. These testcases require chip I/O pin data plus SRL initialisation data as input, and generate "traces", consisting of chip I/O pin data plus SRL status data (signatures) as output. These "functional patterns" are used for the Functional Pattern test on the VLSI tester (and also for the System bring-up, see "7.3 System Bring-Up and Test" on page 316). To be useful on the VLSI tester they have to be converted to the test vector format acceptable by the VLSI Tester.

The tester has to emulate the system environment for the chip under test. This means that both stimulation pulses (input) and output data strobing must be performed by the tester as if the chip was surrounded by the other parts of the chip set and interconnected with them.

Besides appearance of positive and negative pulses in the simulation traces, the pulse duration itself and the sensing of logic High, Low, and Tristate levels at the chip outputs have to be coded into the test vectors. Additionally, one of the appropriate timing sets has to be selected for each test vector, depending on the command that is being executed.

As the environment conditions for a module on the tester differs from the system environment, implementation of voltage levels and timing edges for real time functional pattern execution requires narrow tolerance limits.

Compared to the relatively slow stuck fault "structural" test (usually 500 ns - 1 µs cycle time), the functional pattern test is executed on modules at system speed and under system voltage conditions. It can find AC type faults that cannot be caught by other tests.

5.2.1.5 Logic Test Equipment

Wafer and module test for the Capitol chip set were executed on a VLSI tester "Advantest T3340".

This test system has 256 I/O channels and applies test vectors to the device under test at a maximum rate of 40 MHz. The input signals for the chip can be composed of 32 programmable timing edges and 24 waveforms. Timing changes from pattern to pattern (on-the-fly) are possible with one out of 16 timing sets. Output sensing is supported by multiple threshold voltages and programmable loads.

The tester is equipped with a module station (for chips mounted on modules) and a wafer probing station. Adaption for different products is done with specific front end adapter boards.

5.2.2 Failure Localization and Characterization

5.2.2.1 Second Metal Test

Once a failure is detected, it has to be localized on the chip. As described in "4.2 Master Image Chip" on page 204, the technology provides a wiring capability on three layers of metal, with the third metal level used for power distribution and signal redistribution. The logic functions can be verified on the VLSI tester after completion of second level metal processing with limited performance and noise immunity. At this point mask verification and detection of potential mask repeaters provide an early feedback to the production line and the design community. The second level test is one of the three tests performed on the VLSI tester. Because third level metal is still missing, the exact location of the fault is possible. The wafers are fixed on a temperature controlled chuck. The chip under test is contacted by a 256 pin peripheral needle probe, which is shown in Figure 171.

Test stimulation for the logic part of the device under test uses the complete set of LSSD patterns generated within the design system. This achieves a test coverage of >99% for Stuck-AT faults. Testing of embedded arrays is described in "4.5.4 Testing of Embedded Arrays" on page 222. Test rates were in the 1 us range. Higher test frequencies are possible, but without third level metal, power noise will increase to unacceptable levels, and produce intermittent fail-

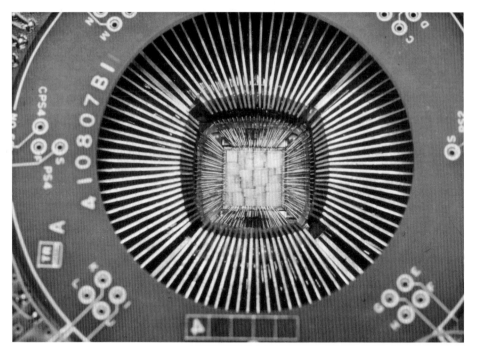

Figure 171. Adapter board with 256 pin contactor probe

ures. The test results are analyzed and the fails separated into random and systematic categories.

If the test sample includes no chip position that completely passes all tests, e.g. during technology bring up or because the sample is too small, the results of successful tester loops (one tester loop = scan-in, master/slave clock cycle, scan-out) are overlaid to predict a composite virtual all-good chip. If systematic fails are detected, the error results of the LSSD pattern are fed into a software diagnostic tool to locate the physical fail area on the tested product. If the area is sufficiently small, the chip can be directly examined by Physical Analysis. Otherwise, if the cone of influence is too large, internal probing may be used for further analysis.

5.2.2.2 Internal Probing Station

In addition to the fail information gathered during second metal test, selected chip-internal logic states are measured with a probe that contacts the 2nd metal signal lines on the chip. Tracing of electrical signals along metal lines and logic circuits is performed while the complete chip is in a dynamic operation mode. The chip is working under fully defined and initialized conditions; signals are observed between sending latch and receiving latch or between input pin and output pin.

The Internal Probing Station used during Capitol chip set development differs much from traditional bench set-ups that bring only small parts of the chip into a working mode, for example a 3-way NAND gate. Therefore, these traditional methods are using only about 3 to 8 needle probes for chip stimulation and sensing.

During Internal Probing

- power and all input stimuli have to be supplied, and output signals sensed, while the logic area of the chip remains uncovered to keep it accessible by a probe.
- The voltage level of a certain net has to be made visible (oscilloscope/E-Beam) without impacting the logic state of this net and without falsifying the test result.
- The VLSI tester has to be attached to the Internal Probing Station, which allows execution of the full pattern range to a large number of I/O pins.

The Internal Probing Station is based on a pneumatic table to absorb vibrations that could remove needles from the chip pads or probe point. A revolver microscope offers a magnification of up to 1000x and a large working distance of 12 mm between microscope lens and chip surface. This permits the use of an active Pico Probe, which is the most important part of the Internal Probing Station. It consists of a 2.0 μm diameter needle connected to an active amplifier. The amplifier has an input capacitance of less than 1 pF. The probe output is connected to a conventional oscilloscope.

The VLSI tester and the Internal Probing Station are electrically interconnected by 50 Ohm coax cables. They are connected to a standard adapter board on the module test station and to the needle probe card at the Internal Probing Station. Due to the capacitive load of these cables, the specified voltage limits of the off-chip drivers sometimes cannot be reached within the standard test cycle of 1 us. This applies especially for the high impedance state, which, in addition to "high" and "low", is a valid and measurable level for the tester. Therefore, the test cycle is relaxed to 3 μs. Due to the LSSD design, the cycle time has no influence on the logical behavior of the chip under test.

On the Internal Probing Station, the chips on the wafer are contacted with the same peripheral needle probe as on the wafer station for the 2nd metal test (see "5.2.1.5 Logic Test Equipment" on page 259). The peripheral placement of the chip I/O pads and contact needles allow contacting of the pads at the chip boundary. Thus it is possible to access the logic area with the Pico Probe. The test vectors can be completely executed in the same way as on the 2nd metal wafer test.

Furthermore, the tester allows specification of a trigger pulse for each individual test vector (input for the VLSI tester). The test pattern set under investigation is applied to the chip in loop mode and the trigger pulse is used to synchronize the oscilloscope with the test vector.

The LSSD design methodology is a prerequisite for this probing method, as the chip is fully initialized after the scan-in section of each tester loop. The subsequent system cycle (master/slave clocking) performs a logic function and stores the resulting data in the latches for the following scan-out and compare section. Therefore, for logic verification via Internal Probing, only those test vectors must be considered that build up the master/slave clock cycle.

The design is supported through rules, which describe the logic books, the connection between them and the physical layout. To make internal probing effective, the output of each logic book must appear on 2nd metal, as 1st metal lines are covered by a passivation layer and therefore cannot be accessed by the Pico Probe. This allows to isolate the physical location of a failure down to one logic book and its stimulating wiring.

5.2.2.3 Fail Locating by Internal Probing

To invoke an internal probing activity on a defect chip, a logical start point must be chosen and its physical location determined. This is supported by a software diagnostic tool. It processes the fail data results from 2nd metal wafer test and returns possible failing wiring nets and their predicted logical levels.

The failing test vectors are executed in a looping mode on the Internal Probing Station and the Pico Probe is positioned according to the layout of the suspected circuits. The values measured are compared with those predicted by the diagnostic tool. The failing location is found, when the measured logic level of a certain circuit does not match the expected one.

Net shorts and metal opens are the two most frequent failure types. A short between two nets results in a level shift in case of opposite values on the shorted lines. An open shows as a solid stuck '0' condition. A diode, connected from each FET gate to substrate ground, floats the remaining wire stub to down level and produces such a stuck fault.

Compared to open nets, shorts are more difficult to analyze because the stuck fault model does not return helpful diagnostic feedback. In this case, the complete logic "cone of influence", in which the failure occurred, has to be pico-probed. The logic paths from the failing chip output pin or from the internal latch are traced backwards through the logic circuits towards the chip input pins or latches where the failing signal may originate.

Facilities are available to show the internal logic state of a net to the operator for any desired test vector. As described earlier, the hardware measurements are compared with the simulation results described in "3.6 Logic Design Verification" on page 184.

5.2.2.4 Performance Verification

The performance of the final hardware is a key item to validate the design goal. Typical path delays represent the design better than speed indicators like oscillators or open loops. Since the delay time for a functional path can be

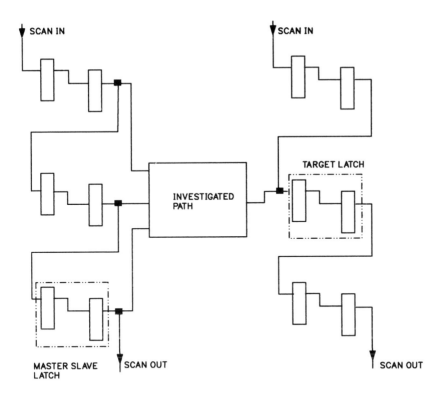

Figure 172. Principle of logic path delay measurement

calculated with the same tools that are used for logic design timing optimization (see "6.5 Physical Design Experience" on page 298), these tools can be verified through hardware measurements. The selected path has to include a representative set of logic books and should be as long as possible to optimize measurement accuracy. The "critical path", i.e. that path which determines the minimum allowed cycle time, seems most suitable.

The principle of a path measurement is shown in Figure 172.

The investigated path is stimulated by scanning the appropriate patterns into the LSSD latches. The signal propagation through the path starts with the last slave clock of the scan-in section. Simultaneously with the scan-in setup, the latch driven by the path is loaded with the complement value to the expected data. By applying the master clock after sufficient time, the target latch will be set.

If the gap between the last slave-clock (which released the data signal) and the master-clock (which captured the propagated signal at the target latch) is large enough, the value in the target latch will be changed. This is verified by a subsequent scan-out and compare operation. If the target latch was changed, the gap between slave and master clock was longer than the time needed for the signal to propagate through the logic path. Therefore, this time gap must be reduced and the complete sequence, (scan-in, slave/master clocking, scan-out) must be repeated.

The path delay is finally determined by the time gap between this slave and master clock pulses when the data of the target latch was not changed. This implies that the pulse propagation through the stimulated path took longer than the delay between the two clock signals. The minimum programmable step size of 125 ps for the clock edge setting allows accurate path delay measurements.

Based on these measurements, hardware performance dependencies, for example delay as a function of temperature, voltage or process parameters, are analyzed. The quality of the logic design delay calculation tools can be judged by comparisons with the simulated path delays.

5.3 Semiconductor Process / Device Design

Karl E. Kroell, Michael Schwartz, Donald Thomas

5.3.1 The Semiconductor Process

The Capitol chip set design is based on a proprietary CMOS process technology which has been developed by the IBM development laboratory in Burlington Vt, USA.

A cross section of the layers vertical to the chip surface is shown in Figure 173. The fabrication of such a structure involves several carefully controlled processing steps, e.g. semiconductor doping by ion implantation, deposition of various insulating and conducting layers, printing of fine patterns into a photoresist film and transferring the image into the layer below by proper etching techniques.

The following gives a simplified description of the wafer processing sequence which is used to achieve the technology depicted in Figure 173. Manufacturing starts with a heavily p-type doped mono crystalline silicon wafer. The good electrical conductivity (0.01 Ohm cm) of this wafer prevents any significant voltage drops within the bulk silicon during electrical operation of the chip. For a CMOS technology this is important in order to suppress the so-called latch-up phenomenon. This is an undesired parasitic bipolar transistor effect which could easily destroy devices and metal lines by excessive currents. The doping of the silicon wafer is too high to place the devices directly into its surface. Therefore a more lightly doped p-type epitaxial silicon layer (12 Ohm cm) is grown on top.

Processing proceeds with the photolithographic patterning and ion implantation of the n-well. This n-type region provides a diffused pocket for the p-channel FETs. Following this, a recessed field oxide is formed by local

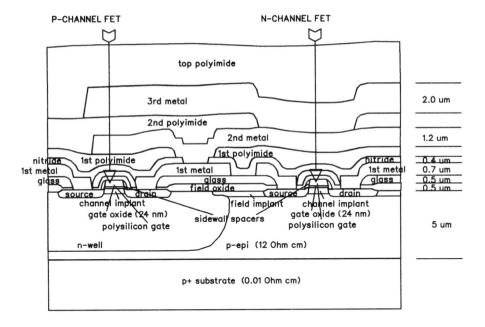

Figure 173. Cross section through the various layers of the process. Approximately drawn to scale

oxidation of the silicon surface. This oxide grows in the field regions between the devices, but not in the device areas themselves, since they are protected by a nitride layer during the growth period. The field region beneath the recessed oxide is also implanted with boron prior to oxide growth to assure proper isolation between the diffusions.

Next, a very thin (24 nm) gate oxide is grown and the channel regions implanted to yield the proper thresholds (see "5.3.3 Electrical Device Properties" on page 268). The gate oxide is covered by an n-doped polysilicon layer. A pattern is etched into this structure to form the FET gates and polysilicon interconnection lines. N-type source/drain regions for the n-channel FETs are implanted, followed by a similar processing step for the p-channel FETs. During implantation the ions can reach the silicon surface only in those areas which are not already covered by a layer. As a result the source/drain edges are self-aligned both to the corresponding edges of field oxide and gate. This self-alignment has two advantages. It leads to closer spacings between adjacent devices. But even more important, it improves the electrical properties of the FETs by reducing the overlap capacitance between gate and source/drain regions. It thus depends upon the lateral component of the two diffusions, the augment of the n-well and n-diffusion masks, and various factors controlling the image size through which the implants are made.

Lateral diffusion of the source/drain edges during process heat cycles would lead to an excessive overlap of the gate unless special precautions are taken.

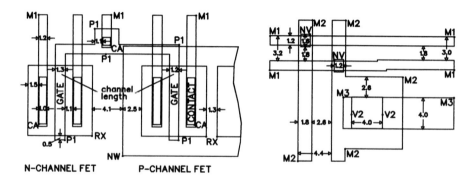

Figure 174. Layout rules for transistors and wiring levels

Prior to the source/drain implantations the sidewalls of the polysilicon gate are coated with oxide spacers. The spacer width is designed to minimize overlap capacitance without affecting proper device operation. Overlap capacitance is key to circuit performance.

After oxide formation on top of the source/drain and gate regions the whole surface is covered by a planarizing glass layer. Contact holes are etched down to the polysilicon lines and to the source and drains. Now the wafer is ready for fabrication of the metal interconnections.

Wafer processing proceeds with the fabrication of three levels of Aluminum wiring layers which are isolated from each other by oxide/nitride or polyimide interlayers. Via holes provide vertical interconnection between Aluminum lines in different wiring planes. The whole chip gets protected by a layer of polyimide. Via holes in this layer lead to Lead/Tin pads which sit directly on top. These final pads serve both as electrical terminals and as soldering points when the chip is mounted on a ceramic carrier.

In total the process utilizes 18 mask levels. 10 masks are used prior to first level metal, the others are required for the subsequent processing. Not all of the masks are equally critical with respect to pattern dimensions and alignment. For example: the BP mask defines a photoresist pattern which blocks out the p-FETs during the n-FET source/drain implant. There is no need to design this pattern with very critical dimensions. Others, such as the poly and contact masks, must be closely controlled.

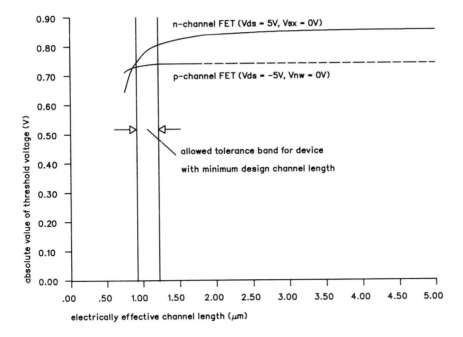

Figure 175. Threshold voltage as a function of electrically effective channel length

5.3.2 Layout Rules

Most of the layers described in the previous section have a thickness of approx. 1 μm or less. This is an important prerequisite to achieving small horizontal structures. The area consumption of device and wiring layouts is largely determined by electrical and photolithographic constraints.

An example for an electrical constraint is punch-through between the n-well and adjacent diffusions of the n-channel FETs. The spacing between these structures has to be large enough to prevent the lateral merging of their junction depletion regions. It thus depends upon the lateral component of the two diffusions, the augment of the n-well and n-diffusion masks, and various factors controlling the image size through which the implants are made.

Specifically photolithographic constraints are:

- Minimum image: this is the minimum dimension which can be reliably printed. The layout rules assume a minimum design image of 1μm for contact holes.

- Image bias: the design image may have a significantly different size on wafer. For example, etching the pattern into the desired layer usually widens the image sizes.

- Image tolerance: a given image in the design can fluctuate in its size on different wafers, from chip to chip and even within a chip. E.g., for the most critical polysilicon level, the layout rules assume a tolerance for the image size of about 0.3 μm.

- Alignment tolerance: during chip fabrication, patterns from different mask levels are aligned to each other. This process is affected by tolerances. The assumption for direct alignment between critical levels is 0.6μm.

All these considerations and tolerances are taken into account in a set of horizontal layout rules. They describe the minimum dimensions which the designer is allowed to use in his layout.

The CMOS process is usually referred to as a 1 μm process. On a reasonably flat surface 1 μm images can be printed into the photoresist film. However, the images usually become wider after etching. Furthermore, during the course of processing the layer thicknesses increase and so does the unevenness of the surfaces. Although the glass and polyimide layers have the desired property to smooth out some of the surface roughness the pattern dimensions usually become wider towards the upper mask levels. As a consequence, while the layout rules are close to 1 μm only for the lower layers, they typically grow in the upper layers.

Figure 174 demonstrates important groundrules. A n-channel/p-channel FET pair is plotted on the left side. Minimum distances and spacings are shown. The minimum design channel lengths of 1.3μm (n-FET) and 1.2μm (p-FET) should not be confused with the electrically effective channel lengths of these devices. From an electrical point of view, the channels are shorter (1.1 μm with a tolerance of +/- 0.3 μm).

The wiring layout rules in the right hand part of Figure 174 allow for metal pitches of 3.2μm (1st metal) and 4.4μm (2nd metal) including the via holes between the two levels. The layout rules for the 3rd level of metal are much wider, since this layer is used only for power distribution and connection to the chip terminal pads.

5.3.3 Electrical Device Properties

The FET devices are designed for 5V power supply operation, imposing several constraints. All breakdown voltages of junctions and gate oxide have to be significantly higher than 5V. This holds also for the FET source drain breakdown, which is due to a parasitic bipolar action. If the channel length drops below a certain level, known as "punch through", the bipolar action produces a parasitic current flow between the source and drain which can detract from circuit performance and increase power consumption.

A second bipolar action which can occur is known as "latchup". It involves a parasitic npn device between the n-diffusion and the n-well, and a pnp device formed by the p-diffusion and the substrate. It is controlled by maintaining proper diffusion spacings and doping levels.

Semiconductor Process / Device Design

The FET threshold voltage is an important parameter for circuit design. It has to be sufficiently high to provide reliable shut-off. The threshold decreases with shorter channel length. This is caused by a short channel effect: With decreasing channel length the source/drain depletion regions interfere more and more with the gate depletion region, lowering threshold voltage. The channel and source/drain doping profiles have been designed to keep the short channel effect tolerable at the shortest allowed effective channel length of $0.8\mu m$. For example, the n-channel FET source/drain regions are made by a double implantation of high concentration arsenic and low concentration phosphorus. The phosphorus implant profile extends beyond the arsenic profile and forms a shallow concentration gradient close to the junction. By this means, the source/drain depletion regions do not extend as much into the p-region and under the gate, and therefore, the short channel effect is reduced. Figure 175 shows the corresponding relation between threshold voltage and channel length both for the p- and n-channel FETs. The double implanted n-FET source/drain structure has an additional advantage: The electric field at the drain side becomes lower and this improves the reliability of the transistor.

It is important for circuit performance to have FETs with optimum transconductance. Towards that goal two parameters, channel length and gate oxide thickness (including their tolerances), have to be minimized. As the channel length is determined by photolithographic patterning of the polysilicon layer, this process has to be controlled carefully. This process achieves an electrically effective channel length of $1.1\mu m$ with a tolerance of $+/- 0.3\mu m$. The 24 nm gate oxide thickness yields a normalized transconductances of $76\mu mhos/V$ for the n-channel and $27\mu mhos/V$ for the p-channel FET, both for low electric fields. The different values of transconductance for the nFETs and pFETs originate from the different mobilities of electrons and holes. (The normalized transconductance is defined as the product of mobility times gate capacitance/area).

Resistances of the various conducting layers also play an important role in determining circuit performance. The sheet resistances for the implanted regions, for the polysilicon layer and the wiring levels are as follows:

Layer	Sheet resistance (Ohm/sq)
N-FET source/drain	36
P-FET source/drain	96
Polysilicon	25
1st metal	0.07
2nd metal	0.044
3rd metal	0.024

Compared to the metal levels the doped semiconductor layers (first three lines of the Table) naturally have a significantly higher sheet resistance. These layers (especially the polysilicon) may still be used for short distance wiring within a circuit, but for long distance interconnections the metal layers are much better suited.

5.4 Failure Analysis

Bernd Garben

5.4.1 Purpose of Failure Analysis

Failure analysis plays a key role in identifying failure mechanisms on VLSI chips. In the course of chip and process development the understanding of failure mechanisms is important both for early detection of design and technology deficiencies affecting the yield or reliability, and to improve the manufacturing process and process control for defect reduction.

Examples for electrical failure mechanisms in the case of CMOS circuits are leakage (shorts) in or between metal layers or polysilicon layers, high resistance (interruptions) in metal or polysilicon layers or at interconnections (via holes, contact holes), gate oxide leakage, source-drain leakage, p-n junction leakage, punch through, latch-up, and threshold voltage (VT) shifts.

It is the aim of failure analysis to determine the physical origin (cause) of a failure. There is a long list of possible causes, e.g. particles, photo defects, mask defects, mask misalignment, geometry of structures on the chip not as specified (metal or polysilicon line widths, size of contact holes etc.), metal interruptions at steep steps, metal spikes, contamination, corrosion, pinholes in the gate oxide or in the insulation between the metal layers, residues (of metal, polysilicon, photo resist etc.), mechanical damage, crystal defects (dislocations, stacking faults, metallic precipitations in the silicon), and finally design errors or weaknesses.

5.4.2 Failure Analysis Strategy and Methods

The starting-point of failure analysis is the electrical test, e.g. a defect monitor test or the functional test of wafers and modules, see "5.2 Test and Characterization" on page 255. The functional tests indicate essentially 4 types of malfunctions on microprocessor chips and modules: leakage between the supply voltage Vdd and ground ("power shorts"), I/O leakage or opens, logic fails, and fails in the embedded memory arrays.

Fails in the embedded arrays are mostly well localized by the bitfail-maps delivered from a functional test. These maps indicate the failing memory cells.

However, the accurate localization of the failing circuits or interconnections on the chip often requires additional efforts, e.g. in the case of leakage between the supply voltage and ground, see "5.4.3.4 Leakage Between Vdd and Ground" on page 274. Some special analysis methods used for defect localization are listed below. For logic fails a software tool was used after the test to identify the failing circuit or at least the failing net, see "5.2.2 Failure Localization and Characterization" on page 259. Since this is very time consuming, failure analysis started with the embedded arrays (Cache and Keystore), and with defect monitors, which are specialized structures that are grouped on the wafers between the microprocessor chips.

The defect monitor tests provide some insight into the electrical failure type (e.g. metal opens, shorts etc.). In contrast to failure analysis this is nondestructive and yields a large data base. However, generally each electrical failure type can be caused by several physical defects. Therefore, a detailed failure analysis on selected wafers is indispensable for a full understanding of the defect mechanisms on defect monitors too.

Moreover, the analysis of logic fails, I/O-fails and power shorts is indispensable and was done in case of systematic fails (e.g. due to mask defects), if a certain type of malfunction accumulated (e.g. due to the physical layout), and/or if the fail occurred in a reliability stress test.

In order to find the physical origin of a failure, a variety of failure analysis methods have to be employed, often in combination with stepwise chip unlayering. The most frequently used methods are:

- failure localization:
  "internal probing" (electrical bench measurements, see "5.2.2.1 Second Metal Test" on page 259), hot spot detection with liquid crystals and voltage contrast scanning electron microscopy (VCSEM)

- microscopy:
  optical microscopy, scanning electron microscopy (SEM) and transmission electron microscopy (TEM)

- micro analysis and surface analysis:
  energy dispersive X-ray analysis (EDX), Auger electron spectroscopy (AES), secondary ion mass spectrometry (SIMS)

- cross sectional SEM and TEM analysis.

These are supplemented by physical and chemical analysis methods for process development and defect analysis on monitor wafers and on wafers which are scrapped during the manufacturing process.

5.4.3 Failure Analysis Examples

5.4.3.1 Particles

Particles can originate from the wafer handling, the manufacturing tools, chemicals, ambient, rinsing water etc. With the decreasing size of circuits and metal lines, even submicron particles can cause a fail. Particle control and reduction requires significant and continuous effort in pilot and manufacturing lines.

A particle can cause various fails, when it simply lies on the wafer during a single process step, even if it is removed during the subsequent process step. E.g. the photo resist exposure, reactive ion etching process or ion implantation can locally be inhibited. Moreover, particles can be embedded on the wafer causing shorts and opens in the metal layers, and gate oxide leakage.

Figure 176 shows a particle, which was embedded in the polyimide insulation between the 2nd and 3rd metal layer. This scanning electron micrograph was taken after partial chip unlayering: the 3rd metal layer and the polyimide were removed (some polyimide "grass-like" residues remained).The particle is approximately 3μm x 6μm large.

Internal probing confirmed that the particle provides a conducting bridge between the two adjacent metal lines causing a logic fail. Micro analysis with EDX showed that the particle consists of iron and oxygen. Neither chromium nor nickel were detected. Because nickel and chromium are present in stainless steel beside iron, this particle can not originate by abrasion from stainless steel tools. The particle may be a rather unique example of iron hydroxide colloids from the DI (deionized) rinsing water.

5.4.3.2 Metal Interruptions at Steep Steps

Figure 177 shows two examples for metal interruptions at steep steps. Photo a) is a scanning electron micrograph of a contact hole which is a hole in the insulation to provide a contact between an aluminum metal line and silicon. The contact holes are etched by a reactive ion etching (dry etching) process into the insulation. The insulation consists of phosphorous -silicate- glass and silicon dioxide. In this example the contact hole edge slope is too steep causing large aluminum grains and enhanced oxidation between the grains. The result is an interruption of the conducting line at the contact hole edge.

Photo b) shows a contact between the first (M1) and the second (M2) metal layer. There the M2-line is interrupted at the right contact edge. Bench measurements indicated a resistance of 165 kΩ, which caused a logic fail. Similar defects have caused reliability stress fails, because the conducting cross section is reduced at steep steps.

Later on steps below metal lines that are too steep or too high were avoided in all metal layers by appropriate ground rules and by a careful control of the edge slope of contact holes during the etching process.

These examples were from the initial stage of the pilot production.

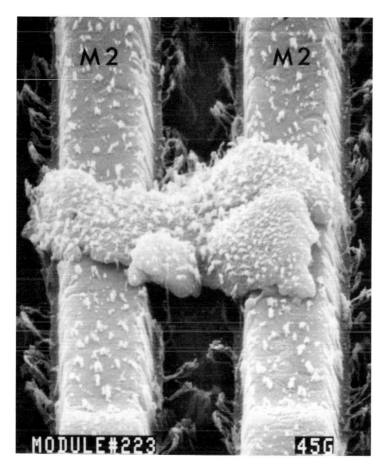

Figure 176. Particle which caused a short between two metal lines

5.4.3.3 Oxide Residues in Contact Holes

In Figure 178 an application of the voltage contrast scanning electron microscopy is shown. A voltage of +5V was supplied to a contact hole chain with a thin tungsten tip in the scanning electron microscope. The contact hole chain is a special defect monitor, where the current flows alternating through the silicon and through short metal stripes. The metal stripes appear dark in Photo a) of Figure 177 up to the point, where the chain is interrupted. Thereby the interruption is exactly localized. Photo b) shows a cross section of the defective area (after the metal and the phosphorous -silicate-glass were etched away). Silicon dioxide residues are clearly visible in the defective contact hole.

During the process development the photo process as well as dry etching process were optimized to ensure that no oxide residues remain in any contact hole.

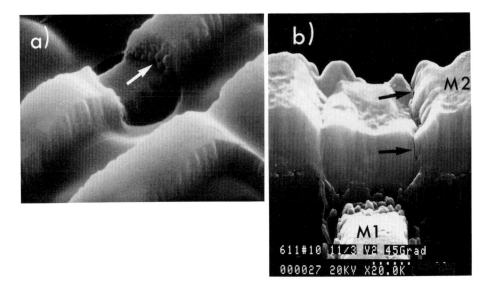

Figure 177. Metal interruptions at steep steps.
   a) metal-silicon contact
   b) contact between the 1st (M1) and 2nd (M2) metal layer

5.4.3.4 Leakage Between Vdd and Ground

Various defects can cause a leakage between the supply voltage Vdd and ground. Often the hot spot detection method allows the localization of the defect which causes the leakage. This method determines the temperature distribution at the chip surface with high sensitivity using liquid crystals.

The hot spot detection method was employed on chips, where the leakage was so large that no functional test could be performed. On these chips hot spots were detected always in the same area in one chip corner. Photo a) in Figure 179 shows a logic circuit after the top two metal layers were removed. The bright horizontal lines belong to the 1st metal layer. Two lines supply Vdd and ground potentials to the p- and n- devices via the contact holes denoted with "1', "2", "3" and "4". A misalignment between the contact holes and the polysilicon lines is clearly visible. Next the 1st metal layer was removed and the silicon in the contact holes etched with sodium hydroxide. Thereby holes in the insulation covering the polysilicon lines were delineated at the contact hole edges (see photo b).

Evidently these holes were etched into the insulation during the contact hole etch process because of the misalignment. As a result there are leakage paths from Vdd to ground via the polysilicon gates (photo a, from 1 to 2 and 3 to 4). The power dissipation in the polysilicon had caused the observed hot spots.

Large variations of the alignment were found over the failing chips. The alignment was critical only in the corner where the hot spots were detected. The reason was that two different photolithographic tools were used to generate the

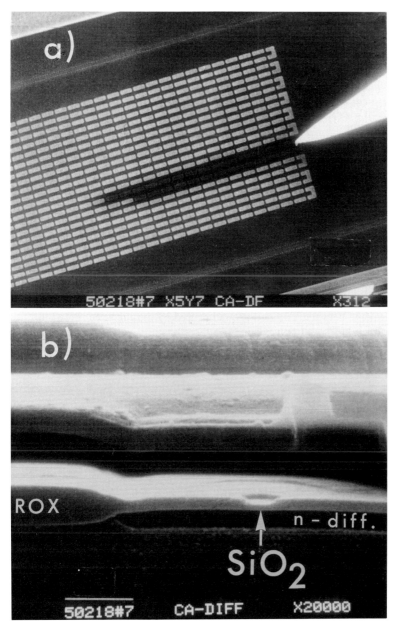

Figure 178. Silicon dioxide residues in contact holes.
a) defect localization with the voltage contrast scanning electron microscopy (VCSEM)
b) cross section

contact holes and polysilicon structures. Small, but different image distortions by two step and repeat cameras caused the critical misalignment just in one

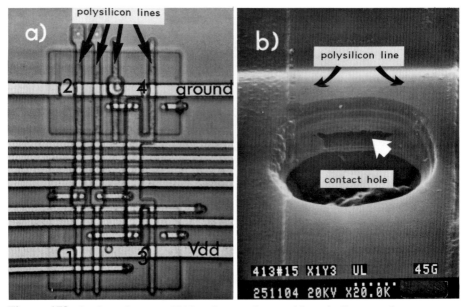

Figure 179. Leakage between Vdd and ground caused by misalignment between contact holes and polysilicon lines.
a) optical micrograph of a logic circuit (2-way AND)
b) scanning electron micrograph showing a hole in the insulation at the contact hole edge

corner of the chips. A very careful adjustment of these tools is necessary to avoid such problems in the case of large chips.

5.4.3.5 Signal to Ground Leakage

The following example demonstrates the importance to tightly controlled geometric dimensions of the structures on the chip.

The electrical test indicated that several tristate drivers on early STC chips reached the logic "1" and "Hz" states, but not the "0". It was already known that the via holes in the polyimide insulation between the 3rd and 2nd metal layers were larger than specified on these chips and had a very shallow edge slope. Therefore failure analysis started with the removal of the 3rd (top) metal layer. Then the via holes at the failing tristate drivers were analyzed with the scanning electron microscope.

There are fissures clearly visible in the polyimide insulation at the via hole edge. They are marked by arrows in Figure 180. The ground potential should be supplied from the 3rd metal through this via hole to a 2nd metal (M2) ground bus only. However, the fissures in the polyimide provide a leakage path between the 3rd metal (ground) and a M2 signal line in the tristate driver running parallel to the M2 ground bus. Therefore this signal line could not reach a potential of +5V and the subsequent inverter output not 0 V.

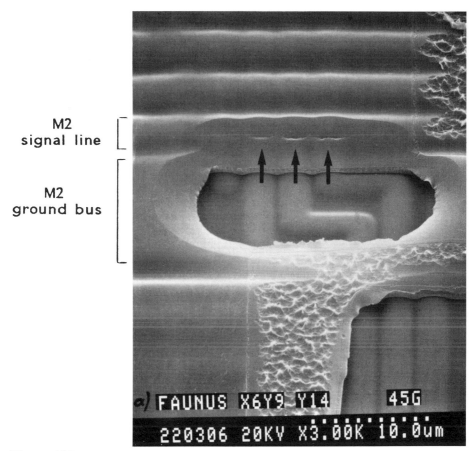

Figure 180. Defective via hole in the polyimide insulation between the 2nd and 3rd metal layer causing a tristate driver fail

The large size and shallow edge slope of the via holes had evidently caused the fissures in the polyimide at the intersections with underlying 1st metal lines. An improved via hole etch process was developed, which allows an excellent control of the via hole size and edge slope. The discussed failure mechanism was eliminated with the new etch process.

5.4.3.6 Source-Drain Leakage

On the first chips, which were delivered by the pilot line, the electrical test indicated fails in defect monitors caused by source-drain leakage. The origin of the leakage was found by the physical failure analysis. Figure 181 shows a transmission electron micrograph of a transistor with high source-drain leakage. Evidently there is an insufficient overlap between the polysilicon gate and the finger-shaped local oxide which limits the transistor channel sideways. As a result there is a constriction of the gate down to 0.4 μm, which causes the leakage (punch through).

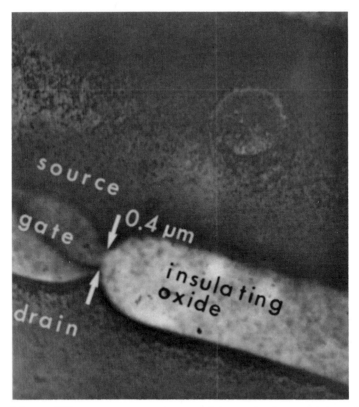

Figure 181. Transmission electron micrograph of a transistor, which failed because of source-drain leakage

A detailed analysis revealed that the "oxide-fingers" were 0.2 μm too short. The reason was a temporary scattered radiation problem with the step and repeat camera.

At the same time, failures of memory cells in the embedded arrays were traced back to the same failure mechanism. It was evident that the cell design was very critical in respect to the overlap between the gate and the local oxide. As a consequence the design of the memory cell was slightly modified. A less critical design helps to achieve higher yields later in manufacturing.

5.4.3.7 Latch-Up

Reliability evaluations during the development of VLSI chips assist in early detection and elimination of reliability related design and technology deficiencies. Part of these evaluations is a voltage screen test, where the voltage supplied to the chip is increased above +5V.

However, at Vdd values of 8.0 V the Cache macro failed on many chips. For all these fails the reason was found to be an inverter in the Cache periphery, which was destroyed by latch-up. (Latch-up denotes an instable condition of a

Failure Analysis 279

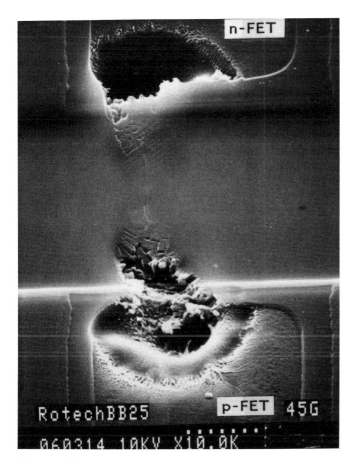

Figure 182. Silicon damage caused by latch-up

CMOS circuit, which occurs when the parasitic bipolar npn and pnp transistors are switched on).

Figure 182 shows the contacts of the p- and n- devices of this inverter after complete unlayering down to the silicon surface. The circuit was evidently thermally destroyed due to the high current flowing through the silicon between the n- and p- device. Thereby the aluminum melted at the contact edges and alloyed with the silicon. The aluminum - silicon compound was etched away during chip unlayering.

A detailed electrical circuit analysis revealed that this inverter was very sensitive to latch-up at elevated supply voltage during switching off (dynamic latch-up triggering effect). As a consequence the inverter was redesigned. With an n-type guard ring between the n- and p- device the latch-up sensitivity was significantly reduced and thereby the reliability of the product improved.

Part 6. Physical Design Tools

6.1 Physical Design Concept

Gunther Koetzle

The need for reduced product development cycles requires an automated VLSI chip design process. It should be as automated as possible, and as customized as necessary to achieve the maximum productivity and earliest product marketing.

The Physical Design Concept both minimizes the design effort, design time, and computer resources, and maximizes flexibility (to support new technologies and circuits), extendibility (to allow CAD for wafer-scale integration), and chip density and performance. The ASIC (Application Specific Integrated Circuit) design system supports multiple part numbers. It provides the user with a given chip image, a circuit library and a set of CAD tools to design the individual VLSI chips.

The existing Engineering Design System (EDS) used within IBM for automated chip physical design was extended to support a hierarchical design approach for VLSI chips. This includes in particular programs for chip partitioning, floor planning and placement, since these areas had unique and new requirements. The hierarchical approach reduces the CAD computer requirements substantially, compared to a "flat" approach, where the total chip logic is placed and wired as one unit (see Figure 183).
"Physical Design" uses the BDL/S logic description, generated by the logic design system as an input (see Figure 95).

As a first step, the total logic is broken into sub-units (regions) of 1K to 3K gates. The most important step for a successful VLSI chip design is floorplanning. There the regions are placed on the chip such that the global nets (long wires) are minimized.

During placement of the regions the area is dynamically assigned, depending on various parameters. One of the key objectives is equal wiring density across the chip. A denser placement is allowed in the peripheral regions due to less global wiring traffic, whereas in the center of the chip more cell locations are unused (depopulated) to provide free wiring channels for the crossing of global wires.

All global nets are considered at the floorplanning phase, and wiring channels are assigned. This allows to brickwall the regions, since all global wires

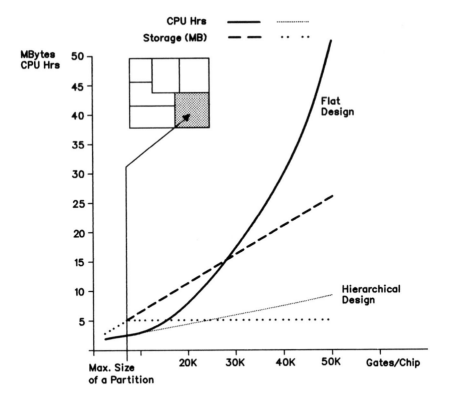

Figure 183. VLSI CAD computer requirements

affecting or crossing a region are known and wiring channels have been assigned.

The regions are now separated and grouped into partitions. A detailed placement and wiring is performed for each partition. The partition to partition connections are controlled to ensure that the connections between neighbouring partitions fit together.

After completion of region wiring, the regions are stitched together by means of a mask data merge (like a puzzle) and the VLSI mask data are complete. A full checking of all process groundrules as well as a logical to physical check is performed to guarantee correct hardware at the first pass.

A timing analysis is done after the floorplanning step with an accuracy of about 5% based on the known global wires and Manhattan distance assumptions inside the regions. Thus, timing induced physical design iterations are prevented. An exact signal delay calculation is done after merging the regions, utilizing all physical parameters.

Plate 5 shows the placement, Plate 6 the wiring of the CPU chip. The MMU chip wiring is shown in Plate 7.

6.2 Hierarchical Physical Design

Uwe Schulz, Klaus Klein

6.2.1 Methodology

The physical implementation of complex and large VLSI-designs requires a Physical Design Methodology that produces chips with high circuit density, allows easy design changes, partitions data into manageable portions and permits a short development cycle. The Capitol chip set has been designed with a hierarchical top-down approach in order to minimize the needed resources, to overcome any limit as to the number of gates and connections, and to gain the necessary flexibility during the design cycle.

The Physical design of the chip utilizes the BDL/S data generated via the tools described in Part 3 as input, and generates mask data that are used to automatically drive the mask generation equipment. It is done in four major groups of steps:

- Partitioning and Floorplanning cuts the design down into regions, places regions on chip, assigns space, and groups regions that form a partition.

- Implantation performs routing of global wires, and the generation of interconnect points.

- Detailed processing extracts, places, wires, and checks the partitions, and generates geometric shapes. This is done a partition at a time.

- Finally, the chip physical design is assembled by merging the partitions and checking the data.

The process flow of these 4 major physical design steps is shown in Figure 184. All steps are performed by a group of CAD programs that are extensions of IBM's Engineering Design System (EDS). These programs perform test data generation as well (see "5.1.3 Structural Test" on page 244). At certain points the physical designer may interact with the execution of these programs. However, the mask generation data are fully machine generated from the BDL/S input file.

6.2.2 Partitioning and Floorplanning

The entire BDL/S design is cut down into smaller pieces, called regions. Pregrouping of logic is accepted, i.e. a defined part of the logic may be divided into a number of regions but will not be mixed with the rest. The subdivision of the BDL/S design into regions is based upon the number of interconnections and the desired sizes (min-max). of the region. Those parts of the logic that have more internal than external connections may form a region. Figure 185 shows the working of the partitioning program.

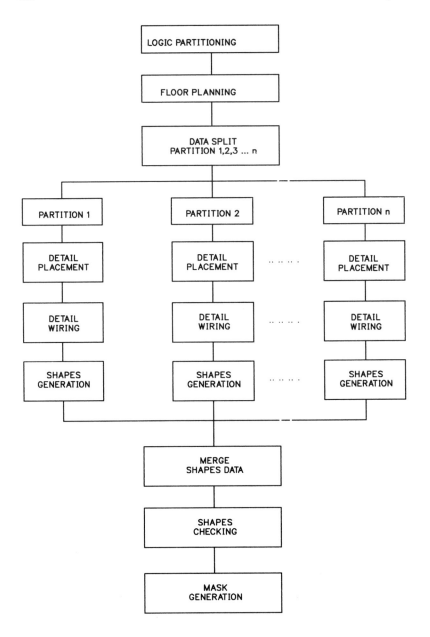

Figure 184. Hierarchical physical design processing flow

After regions have been found and identified, they are treated like junctions and get positioned on the chip depending upon their interconnections. For this, the sum of the length of all interconnections is minimized, using the rule: the heavier the "traffic" the more adjacent the positioning. This lowers the space requirements for global wires. A different pattern of positions is honoured if required for timing reasons. Thereafter, area is assigned dynamically to the

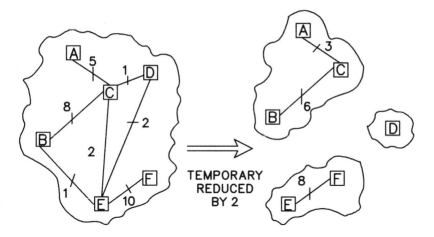

Figure 185. The sum of all interconnections between the units of logic is continuously lowered until islands -regions- emerge (sum = 0). Regions of a size below min get combined with others automatically. The outer boundaries of pre-grouped logic is untouched.

individual regions, taking into account the number of circuits, pins, and connections. The aspect ratio (x,y) of 1 is tried to achieve wherever possible. This results in a floor plan as shown in Figure 186.

Multiple regions are now grouped into partitions. A partition can be as large as the chip or as small as a region. Criteria for selecting the partition size are computer runtime, program limitations, and computer resources. By means of partitioning any chip size can be processed. This results in the floor plan shown in Figure 190.

6.2.3 Implantation

This step generates interconnect-pins (IC-pins) at the borders of the partitions and supplies an adjustment figure as to the space requirement for global wires. In order to have additional space for signals that enter and leave the partitions, a set of interconnect-areas (ic-areas) is defined. With the help of a pseudo-routing across these ic-areas (either automatically or manually) an IC-pin is generated for each signal (see Figure 191). This IC-pin acts like a place-holder but has yet no fixed location in that ic-area. Routing is done by bundles of nets, i.e. all signals that have equal sources and sinks get bundled.

When all nets are routed on the shortest possible path, some ic-areas may have more ic-pins than others. Therefore, ic-area size has to be controlled tightly, which, in turn, may result in longer routes.

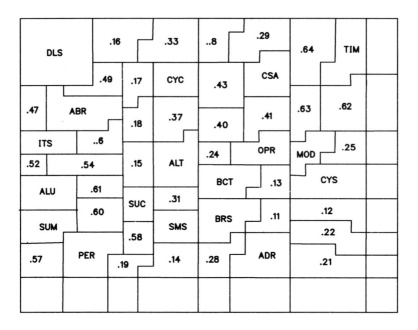

Figure 186. Positioning of regions on chip based upon the number of interconnections and area assignment. Goal of aspect ratio is 1.

6.2.4 Detailed Processing

All data to process a partition are now available. Partitioning in conjunction with floorplanning supplies the content and space requirement. During implantation on a chip, ic-areas are defined and ic-pins, which are provided by ic-circuits, are generated. In addition, an adjustment figure as to space for global wires is calculated. The process steps discussed so far are shown in Figure 187. Subsequent steptps are executed a partition at a time.

Hierarchical Physical Design

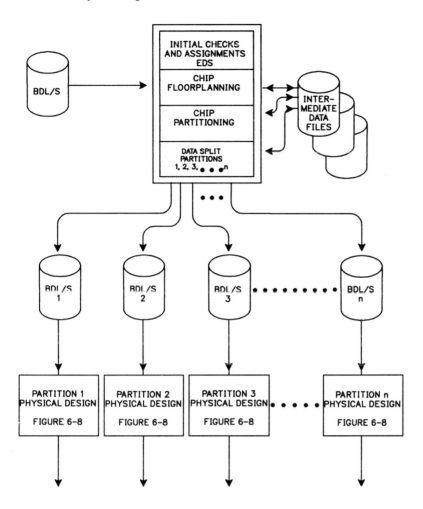

Figure 187. Partitioning process steps

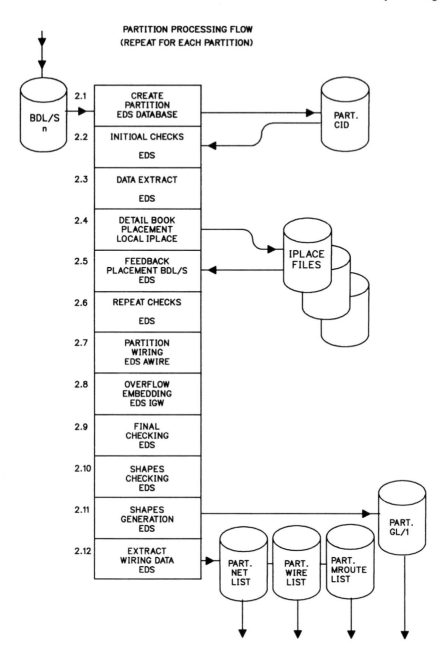

Figure 188. Partition processing

Hierarchical Physical Design

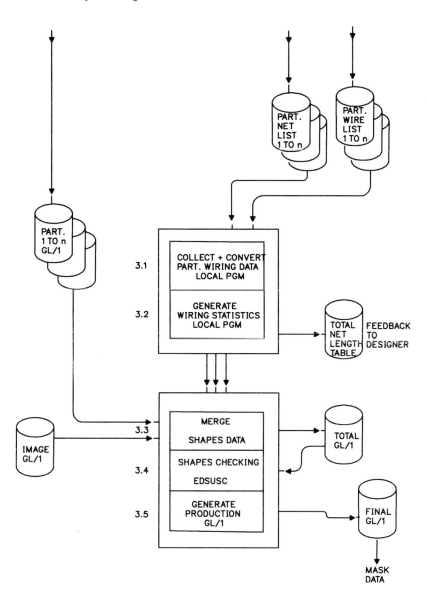

Figure 189. Chip assembly processing

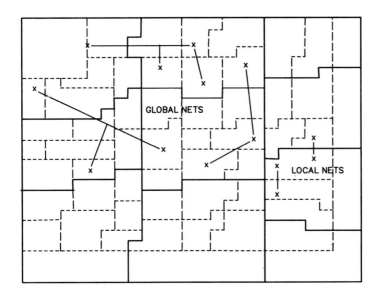

Figure 190. Grouping of regions to form partitions

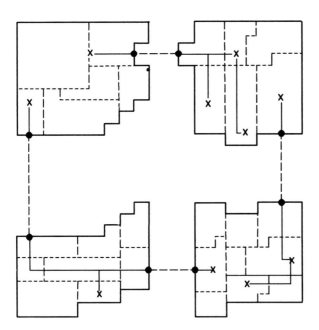

Figure 191. Generating interconnect-pins (ic-pins) by pseudo-routing all nets across pre-defined ic-areas. The ic-pins are place-holders within the ic-areas and receive fixed locations when affected partitions get processed. In addition, space adjustment for global wires is done.

Hierarchical Physical Design   291

When processing the very first partition, all ic-circuits (-pins) 'float' within their ic-areas, i.e. they are treated like any other circuit of the logic and get assigned a fixed and best suitable location while all circuits of a partition are placed. After processing a particular partition, all ic-circuits of that partition have fixed locations. Therefore, all ic-circuits of adjacent partitions that have been processed before, get selected and treated as pre-placements (they are also used for checking).

All logic circuits within a partition get placed during processing that partition. Should a partition consist of more than one region, the circuits of a region will stay within the boundaries of that region. This technique allows to pre-place a group of circuits. It allows to have different placement densities that result in a better distribution of all connections. For example, the middle of a partition will have a lower density of local, and a higher density of global wires, while at the edge of a partition the opposite relation will hold true. Control of the densities and the ability of the placement program "not-to-cluster" improves the wireability and allows an optimum usage of the chip area.

Spare circuits for engineering changes get placed into free slots after the placement of the circuits of the partition has been completed.

A partition is a "self-contained" piece, i.e. it contains not only the circuitry and internal -local- wiring but also all global connections that enter, leave and cross the partition and are part of the final chip. The next steps are wiring, checking, and shapes generation.

A standard chip wiring program is used to automatically complete the connections of all circuits, transfer circuits, and embedded macros as usual. For very dense parts a small number of connections may be left as overflows. In this case an interactive tool is used to complete the remaining connections on the screen.

After wiring is completed and all final checks are successfully processed, the shapes for the production masks are generated. The processing steps unique for each partition are shown in Figure 188.

6.2.5 Chip Assembly

A final merge run collects the shape data of all partitions and puts them together, each partition having its own X and Y offset according to the floor plan. On the partition boundaries the transfer books representing the ic-pins of adjacent partitions are overlaid. Because they represent wiring channel crosspoints, no additional wiring is required to join global wire segments. The wiring pieces of the puzzle snap together and complete nets are formed matching the overall logic structure.

An extensive final checking of the resulting shapes data completes the physical design process. The partition data are checked against the total image data. In particular, the connection data for signals crossing partition boundaries are checked for completeness and the partition boundaries are checked for overlap conflicts. The chip assembly processing flow is shown in Figure 189.

6.3 Hierarchical Layout and Checking

Kurt Pollmann, Rainer Zühlke

6.3.1 Chip Layout

The VLSI chip images are built from a few building elements: A horizontal I/O cell, a vertical I/O cell and an 'image module' for the internal chip. Figure 192 shows that this image module is 1 horizontal I/O position high and 2 vertical I/O positions wide. It contains 40 positions for the placement of logic books. The power distribution is contained in this module. Any size of chips could be constructed from these elements. The 12.7 mm × 12.7 mm chip size has 38x65 = 2470 modules; the smaller 9.4 mm × 9.4 mm STC chip has an 24x41 = 984 array. This regularity allows to generate shapes of books and wires on smaller working images and move them correctly on the real chip. The working chips must support the largest partition and not exceed the limits of the physical design system, for example the wiring program limits.

The logic books are placed on legal chip positions only. They correctly connect to the power busses and the n-well of the chip image.

The array macros on the MMU chip (cache, key store, cache directory, translation-lookaside buffer) use custom silicon and power distribution. They are placed in chip areas, where the image modules left an empty hole for them. The array macros must provide a "power ring" cell, which connects to both the chip image (power + n-well) and to the array macro internals.

The array macros and the image module contain shapes for possible via holes between second and third metal.

6.3.2 Chip Merge and Final Data Generation

Design automation software merges the shapes of all partitions together with a chip image that has the correct cutouts for the array macros. Now the metal 2 / metal 3 via holes are generated. A program checks which possible via shapes fits correctly under the metal 3 shapes of the image. These shapes are converted to production level. For the final via generation only the metal 3 layer and the possible via holes are used. No excessive data volume is necessary to check from third metal to silicon.

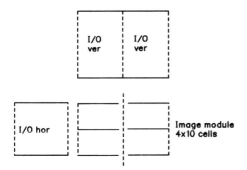

Figure 192. HPCMOS chip image elements

6.3.3 Checking

Groundrules for all books and array macros are verified before they are stored in a rules dataset for controlled usage.

On a mini-image with the size of few cells, all books are placed and wired. This small amount of data is checked down to the silicon for all rules. For this, an USC (Universal Shapes Checking) program and a verified checking program for the CMOS technology was written. This rules verification guarantees that books do not conflict with each other. All wiring is done in the metal layers only. The wiring test guarantees that wiring and via holes shapes create no problem with the silicon shapes. After passing all program runs, the complete chip is "correct by design".

The shapes generated for a partition are checked physical to logical. The nets between all book pins are checked for opens and shorts.

During the final chip check the hierarchy of the array macros is used to hide the large amount of data required for the arrays macros. Only the power ring cell and the array macro pins show up for checking. This guarantees the correct assembling of the power distribution and the logic nets.

For the physical to logic check all first and second metal shapes and all via holes are used. This means correct nets and a continuous power distribution without shorts. If no open or shorted nets occur during these runs, the partitions are correctly placed and merged.

Data volume and run time problems are inherent problems in the hierarchical layout and checking software. The following measures contain them:

- Book and mini-image checks account for all silicon problems.

- Physical to logical check of a partition guarantees correct placement and wiring.

- Physical to logical check of the final chip needs the shapes of 3 layers only and can suppress the internal array macro data.

6.4 Delay Calculator and Timing Analysis

Peter Hans Roth, Hans-Günter Bilger

6.4.1 Circuit Delay

When introducing a new technology, the transient behaviour of the circuits available in the circuit library have to be established early in the design cycle. The logic designer needs to know the performance of the critical path (the longest path in the machine, and therefore the path that determines the machine cycle). For this calculations the delay of all the circuits offered in the circuit library have to be known as early as possible in the design cycle (see "3.5 Timing Analysis and Verification" on page 177). Figure 193 shows a 2-way AND INVERT (2WAI) circuit in CMOS technology. The input A1 of the circuit is assumed to be in the "high" state. The other input is used as the switching input by applying the input voltage as shown in Figure 194. The corresponding output voltage of the circuit is Vout. C is a lumped capacity representing the output load i.e. wiring and fan-in capacitance of the following circuit(s). Figure 194 shows the transient behaviour of the input and output voltage of the 2WAI circuit. It also shows the four delay times which are to be calculated.

Trise : slope of the up-going edge at the output (10% to 90%).

Tfall : slope of the down-going edge at the output (10% to 90%).

Ton : delay between the up-going slope at the input and the corresponding slope at the output at 50% of the total swing.

Toff : delay between the down-going slope at the input and the corresponding slope at the output at 50% of the total swing.

Ton and Toff are strictly related to the input voltage (i.e. the **up**-going input slope always defines Ton to the corresponding output slope).

At an inverting circuit therefore Trise at the output depends on Tfall at the input and vice versa.

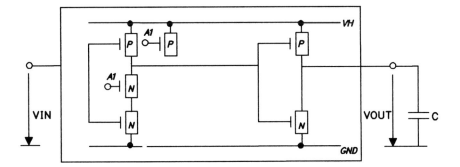

Figure 193. 2-Way NAND circuit

6.4.2 Calculation Method and Simulation

From the geometric design data of every circuit a corresponding circuit model is derived by a program. This circuit model represents the equivalent circuit with all its parasitic elements.

A separate circuit analysis program allows together with this circuit model to calculate and simulate the transient circuit behaviour. For a discussion of circuit analysis programs see [SPIRO]. Figure 195 shows a model that allows to generate a realistic pulse, which will be applied to the input of the circuit under calculation. The source represents a simulated impulse calculated by the circuit analysis program. This impulse is used as input to the 2 way AND INVERT circuit (2WAI).

The output of this circuit generates the desired pulse by measuring the voltage (E1) over a current source I1 = 0. To simulate a realistic environment, another 2WAI acts as a load.

In order to vary the slopes of the leading and falling edges of the pulse, the 2WAI output can be loaded with five different capacitances (0...2.5pF). It is represented by the variable capacity Crf in Figure 195. This pulse (i.e. the voltage E1) is used as input pulse for the circuits under calculation.

Figure 196 shows an example with seven noninverting books. This calculation does not require loading the pulse generation circuitry, because the input pulse E1 is measured across the current source I1 = 0 in Figure 195.

Every circuit under calculation can be loaded with nine (0...10pF) different capacitive loads (C1 in Figure 196) at the output in order to simulate different circuit loadings.

The delay of the circuits is simulated with five different input slopes and nine different output loadings. This yields a total of 45 reruns per circuit.

Four delay values (Trise, Tfall, Ton, Toff) are calculated, so that a total of 180 timing values will be obtained for every circuit.

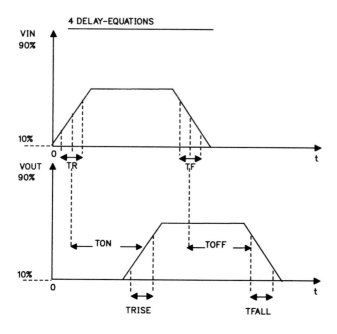

Figure 194. Transient behavior of 2-way NAND circuit

6.4.3 Fitting Method (Least Square Fit)

The delay behaviour can be described by a set of four standardized delay equations, e.g. for an inverting circuit:

```
Trise = (K1  + K2*C)*Tf  + K3*C**2  + K4*C  + K5
Tfall = (K6  + K7*C)*Tr  + K8*C**2  + K9*C  + K10
Ton   = (K11 + K12*C)*Tr + K13*C**2 + K14*C + K15
Toff  = (K16 + K17*C)*Tr + K18*C**2 + K19*C + K20
```

    C : capacitive circuit load in pF
    Tf: slope of the up-going edge at the input
    Tr: slope of the down-going edge at the input

The coefficients K1...K20 are calculated with a least square fit method. depending on the variables C and Tr or Tf.

As an example, the calculated set of delay equation for the 2 WAI is shown below:

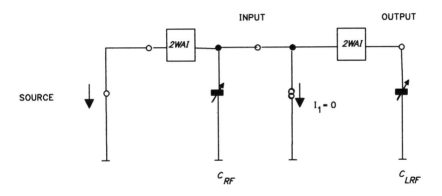

Figure 195. Input pulse generation

```
Trise=(0.218 + -.018*C)Tf + 0.006*C**2 + 4.280*C + 0.379
Tfall=(0.189 + -.014*C)Tr + 0.009*C**2 + 3.168*C + 0.217
Ton  =(0.017 + 0.014*C)Tr + -.015*C**2 + 1.700*C + 0.252
Toff =(0.188 + 0.011*C)Tf + -.010*C**2 + 2.024*C + 0.310
```

Environment Conditions are: Vh = 4.5 V
Temp. = 85 Degree C

Process Parameters: 1.5 Sigma (worst case)

C : capacitive circuit load in pF
Tf: slope of the up-going edge at the input
Tr: slope of the down-going edge at the input

The models in the circuit analysis program can be adapted to the applied temperature and voltage. The line parameters can be set to worst or best or nominal case in order to get worst or best or nominal delay equations.

The actual parameters of a chip are obtained as described in "5.2.1.2. Process Parameter Test" on page 257. These data allow to generate a set of actual delay equations corresponding to the chip under measurement.

Using this approach, worst and best case behaviour of the used circuits can be obtained in a very early state of the design phase. Once real hardware is available, the delay equations can be quickly and easily be adjusted by taking the original line parameters into consideration and by initialising the circuit models with this actual data. This guarantees a permanent feedback between the hardware and the calculated delay equations. For the Capitol chip set the correlation between calculated and measured delays turned out to be within five percent.

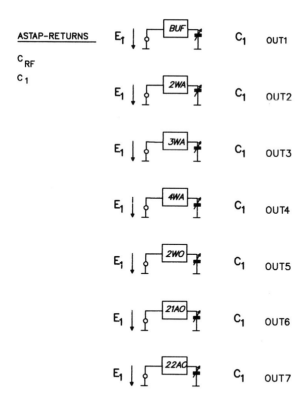

Figure 196. Circuit sample

6.5 Physical Design Experience

Gunther Koetzle

6.5.1 Master Image Development

Various groundrule changes had to be implemented during the Capitol development process. The structured hardware approach (sub-circuit elements) permitted an implementation time for each change of 3 days. Most of it was required to update the rules system.

Three minor errors were found during testsite verification, caused by incomplete checking tools at testsite design completion. No problems due to circuit or physical were encountered on the first silicon CPU and MMU chips. Due to extensive electrical analysis and modelling, the performance of all circuits met the specification (e.g. 2WAI 0.8 ns typical at 0.4 pF). The special design effort taken in the areas of chip power distribution and off-chip driver design resulted in the absence of any noise problem. The chip internal noise and the simultaneous switching noise was measured within the predicted and specified range of less than 600 mV.

Due to the use of LSSD (Level Sensitive Scan Design) groundrules for testing - including all embedded arrays - a testcoverage > 99% was achieved for the processor chips.

6.5.2 Physical Design

The physical design system turned out to be already bug free and fully operational during the test chip design. As expected, the 60k gates (75k cells) of the CPU chip are wirable on the 12.7 mmsq chip image, for a mixture of control logic and data-flow.

The physical turn-around-time, from BDL/S input to checked mask data, was 60 working days target for YE '87 is 40 working days.

The CAD computer requirements for the physical design of the 12.7 mm CPU chip was an average 12 hrs time on an IBM 3081 system, using 5MB of real main storage.

6.5.3 Hardware Bring-Up

Four features implemented to support the bring-up of the first fully functional Capitol chip set system proved to be very helpful:

- An engineering change capability for logic corrections by means of unused spare logic cells distributed over the chip, as part of the unused (depopulated) silicon area. Those spare cells are connected to the power busses. If required, the tied inputs are activated by a metal change only. This allows to perform minor logic changes up to a few days prior to mask data completion. In addition the turn-around-time for wafer processing in case of a logic error correction, is reduced compared to a gate array type, since the back-end of the process (metal) has to be changed only.

- The chip image is designed to be fully functional after second level metal with metal 2 probe pads at the periphery for all signals and power. Thus chip internal signal probing can be done for failure analysis - which is not possible after metal 3, because the M3 power buses and I/O redistribution cover the signal lines in M1 and M2.

- "Probe cells", containing a wider M2 stripe for proper contacting of the pico probes, are provided to the logic design team. They may connect those cells to any net that is considered of interest for observation. The probe-cells are automatically placed and wired.

- To ease the orientation on the VLSI chips for visual inspection and analysis, coordinates are coded into the edges of the M2 power buses to identify any group of 44 cells on the chip.

The logic design team made multiple use of the "metal engineering change" capability to implement logic corrections after start of physical design or to correct logic errors found after hardware availability during system bring-up.

6.5.4 Lessons Learned

Some of the additional experiences gained from the Capitol chip set design are:

- CMOS VLSI circuit speed is much more limited by noise than by driving capability.

- Complex logic cells might ease the logic design but do not necessarily give better VLSI chip performance and/or density. The density advantage of custom macros for logic is lost when the macros are implemented on a VLSI chip due to the impacts/restrictions on global wiring. In NMOS technology the very limited driving capability asked for compaction in macros. These limitations are not as valid for CMOS technologies

- Maximum density on a VLSI chip is not achieved by simply minimizing the circuit count. Since VLSI chips are wiring limited, the better solution for density is the reduction of global wires - even at the expense of additional circuits.

- Using some fraction of circuits for test and RAS (Reliability, Availability, Serviceability) is a good investment.

Part 7. System Implementation

7.1 ES/9370 System Overview

The IBM ES/9370 model 30 and 50 machines were the first users of the Capitol chip set. They differ among others in the machine cycle time (80 ns for the model 30 and 62.5 ns for the model 50). Other realizations are the IBM 3092 Processor Controller Models 004 and 005, and the IBM ES/9370 model 25. In the ES/9370 system, the CPU, main store, I/O bus, input/output controllers, as well as tape and disk devices all mount into a single, or multiple, 19"rack. Figure 197 provides a *logical* overview of the ES/9370 system.

The ES/9370 models 30 and 50 are mounted in a card enclosure, which holds the processor logic, the storage, and the I/O controller cards. The processor itself requires 3 or 4 cards. One card, the CPU card, contains the basic Capitol chip set (see Figure 153). One or two cards contain 4, 8 or 16 MByte main memory. One card contains 2 BCU modules ("2.7.3 BCU Chip" on page 110) and the EPROM ("2.2.2 Block Diagram Description" on page 20). The card enclosure has an additional 10 slots for I/O controller cards. (An I/O controller can consist of more than one card, depending on its type.)

I/O controllers are attached to the ES/9370 Processor through internal I/O busses (see Figure 198). I/O bus, I/O controller, and attached devices appear as channel-attached I/O to the operating system. Multiple I/O buses operate independently of each other. A Service Processor attaches direclty to the CPU card via the support bus.

The internal I/O buses attach to the Processor Bus as shown in Figure 71. The ES/9370 model 50 differs from the model 30 in so far as it attaches 4 I/O busses via 4 BCU chips. All I/O controllers are intelligent and contain their own microprocessor (mostly Motorola 68000). Their R/W memory is loaded from a DASD (hard disk) and is initialised as part of the Initial Microprogram Load (IML) process. I/O controllers perform the combined function of a S/370 channel and control unit. I/O controllers communicate with the CPU and MMU chips, with main storage and with non-S/370 memory via the BBA chips described in "2.7 Bus Interface Chips" on page 108. For data transfer operations, the corresponding I/O controller serves as the I/O bus master, while the Capitol chip Set BBA chips have only a slave function. The S/370 I/O instructions like Start I/O, Halt I/O, Test Channel etc. are implemented via cooperative microcode processes running in both the CPU chip and the corresponding I/O controller. The S/370 channel and control unit functions are

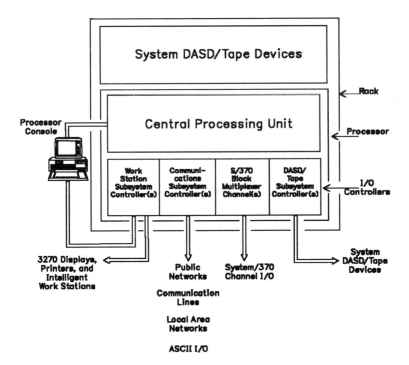

Figure 197. Overview of system components

implemented through microcode in the I/O controller. Computer systems may implement I/O functions with the help of "I/O driver" routines executed by the CPU (usually part of the operating system Kernel) or alternatively via channel programs executed by an intelligent I/O controller. The S/370 architecture specifies the second approach (channel I/O). The performance of a computer system is usually determined by 3 factors: CPU speed, main store (and cache) access and I/O performance. The S/370 channel architecture minimizes the time the CPU spends on I/O control function. This guarantees superior CPU response time in environments with heavy I/O traffic. The Capitol chip set based IBM ES/9370 model 30 and 50 machines share this advantage with other S/370 implementations.

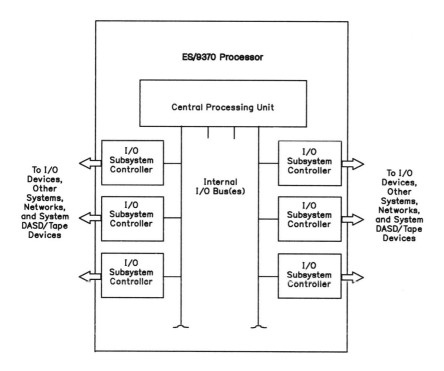

Figure 198. Basic IBM ES/9370 processor structure

7.2 High Level Microprogramming in I370

Jürgen Märgner, Hartmut Schwermer

7.2.1 Overview

Traditionally, a S/370 CPU consists of hardware and microcode. Hardware is used for the implementation of the performance critical S/370 instructions and microcode for the remainder of the S/370 functions.

The Capitol chip set adds a new dimension to microprogramming: the I370 mode. In the IBM ES/9370 Model 30 and Model 50 CPUs a processor kernel of hardware and microcode operates in two modes: S/370 mode, which supports the majority of the S/370 instructions, and I370 mode. Additional code executing in I370 mode utilizes the kernel functions to implement the remaining parts of the S/370 architecture. The CPU either executes instructions in S/370 or in I370 mode. Normally it executes in S/370 mode; it only enters I370 mode to execute S/370 functions implemented in I370 code.

I370 mode is a subset of the S/370 architecture and has special instructions for access to resources that are external to the CPU. Event handling requires interaction with such resources through "control spaces", using special I370 instructions. The control space concept is the main concept that characterizes I370 mode and distinguishes it from the S/370 architecture. This concept is a variation of the well-known "Memory Mapped I/O" concept, which has been applied in various microcomputer systems.

7.2.2 Concepts and Facilities

7.2.2.1 Processor Structure

A conventional S/370 processor consists of hardware and microcode as shown in Figure 199 Hardware interprets the performance critical S/370 instructions and microcode is used to execute the remaining functions.

A detailed analysis of the microcode as illustrated in Figure 200 shows that about 15% of the microcode, measured in lines of code, is simple, straightforward code and is utilized for performance critical functions. The remaining 85% are complex and not performance critical functions. The major problem of traditional microprogramming becomes apparent when locking at Figure 200: microcode is normally implemented using assembler languages, causing a major productivity and quality problem, especially for the 85% part. I370 mode, however, allows the implementation of the 85% part in a more structured and productive way. An example of an I370 program coded in a high-level language is shown in Figure 203.

The introduction of I370 into the conventional processor structure leads to a revised structure, which is illustrated in Figure 201. In this structure a kernel of hardware and microcode supports two modes: A partial S/370 mode, which supports the majority of the S/370 instructions, and I370 mode, which utilizes the kernel functions to implement the remaining part of the S/370 architecture.

The hardware and the 15% microcode part, written in assembler language, account for the majority of the S/370 instructions. The remaining 85% are implemented in I370 code which can be implemented using existing high level languages instead of assembler language. Consequently, as shown in Figure 202 the 85% part can be implemented in a more structured and productive way. An example of an I370 program coded in a high-level language is shown in Figure 203.

The difficult question to be answered is: Which S/370 functions are in the 85% part? During the preceding microcode analysis they have been identified to be:

- I/O Control: instructions, control, and interruptions.
- System Control: IPL, resets, machine check handling, RAS, and operator control functions.
- All processor control functions such as initial microprogram load (IML) or the CPU to Service Processor communication (for those configurations that use a service processor).

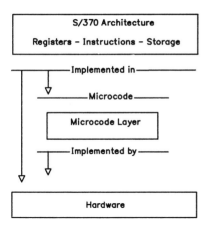

Figure 199. Conventional S/370 processor

Consequently, the S/370 part of the processor kernel and the S/370 part of I370 mode must provide all S/370 functions, except the ones identified above. The kernel functions unique to I370 mode, which are not part of the S/370 architecture, are defined below.

7.2.2.2 Instruction Interpretation

In the conventional processor structure without I370 mode, S/370 instruction interpretation starts in hardware. If the instruction cannot be interpreted in hardware, control is passed to the microcode. If the microcode cannot perform the interpretation, a S/370 program interrupt is generated.

In the processor structure with I370 mode, however, if the microcode cannot perform the interpretation, instead of generating a S/370 interrupt, control passes to the I370 code. A S/370 program interrupt is generated only if this layer cannot interpret the instruction.

The important concept is that I370 code always gets control in case S/370 instruction interpretation cannot be performed successfully by either hardware or microcode.

7.2.2.3 Control Spaces and Associated Instructions

Control spaces are linear byte address spaces. They can be accessed with control space I370 instructions. These instructions are shown in Figure 204. These instructions are unique to I370 mode and do not exist in the S/370 architecture. Contents of control spaces can be loaded into general registers with the Read Control Space instruction. Alternatively, the contents of control spaces can be changed with such instructions as And Control Space, Exclusive Or Control Space, Or Control Space, and Write Control Space.

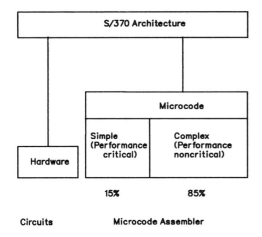

Figure 200. Microcode analysis without I370

All resources external to the CPU and main storage are mapped into control spaces. Control of these resources is provided by reading and writing associated control space locations. In addition, a CPU external event may asynchronously cause a change to the contents of one or more associated control space locations. A control space is called event pending, if one or more bits in word 0 are set to one.

Two types of control spaces are distinguished: architected and implementation specific control spaces. The architected control spaces are listed in Figure 205. The implementation specific control spaces are described in "ES/9370 Realization".

- The Event Control Space (ECS)
  enables efficient event recognition and reaction. The ECS is one fullword in length. The left halfword is dedicated to the architected control spaces. The right halfword is dedicated to the implementation specific control spaces. Each bit identifies exactly one control space. The value of each bit is always the OR of bits 0-31 of word 0 of the associated control space. If any ECS bit becomes one, I370 mode is entered as described under " 7.2.2.4 Mode Control and Associated Instructions" on page 307.

- The S/370 Control Space (SCS)
  provides access to S/370 specific status information. The SCS becomes event pending, if a condition occurs during S/370 instruction interpretation that requires I370 mode invocation to complete the event causing S/370 function. In this case the SCS also contains all the information required to complete the S/370 function. It is 29 fullwords in length.

- The Timer Control Space (TCS)
  provides access to I370 unique timing facilities. The TCS defines two facilities: an interval timer facility, which is conceptually equivalent to the

High Level Microprogramming in I370

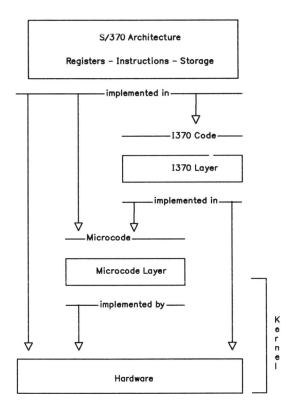

Figure 201. A S/370 processor with I370

S/370 interval timer, and a time counting facility, which is conceptually equivalent to the S/370 time-of-day TOD clock. The TCS becomes event pending whenever the interval timer value is decremented from a positive or zero number to a negative number. The time counting facility does not cause the TCS to become event pending. The TCS is 4 fullwords in length.

- The Programmable Control Space (PCS)
  allows to raise events explicitly with I370 control space instructions. It is one fullword in length. The specific meaning of all PCS bits is implementation defined.

7.2.2.4 Mode Control and Associated Instructions

All standard S/370 general, control, decimal, and floating-point instructions are available in I370 mode. Figure 204 lists the control instructions that are unique to I370 mode.

The I370 unique control instructions are required to support I370 mode exit to S/370 mode, and to enable general register save and restore and program resumption in I370 mode without requiring a base register.

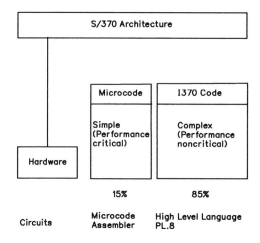

Figure 202. Microcode analysis with I370

I370 mode cannot be entered under program control. It is entered only -

- if the CPU is in S/370 mode and the Event Control Space becomes event pending, or
- while a system start-up or a S/370 reset function is performed.

I370 mode exit is always performed under program control by executing the control instruction Leave I370•Mode (LIM).

7.2.3 ES/9370 Realization

Figure 206 shows the Capitol chip set in a 9370 configuration. Please note the use of a service processor, and that I370 is an integral part of the CPU. The physical code and data residence is certainly processor storage as it is for regular S/370 programs. This particular storage area, non-S/370 memory, is not available for customer programs or data.

Under supervision of the service processor, I370 code controls the system from the very first IML (Initial Microprogram Load) steps until the system is up and running. After IML is complete, other processor functions implemented in I370 are made available to the system such as the support for S/370 programs communicating with I/O devices. I370 will assist during setup and termination of each I/O operation.

```
Dispatcher: PROCEDURE;
DECLARE ECS      LITERALLY('0'), /* Event-Control-Space    */
   SCS    LITERALLY('1'), /* S/370-Control-Space    */
   TCS    LITERALLY('2'), /* Timer-Control-Space    */
   PCS    LITERALLY('3'), /* Programmable-Control-Space  */
   PU_ACS LITERALLY('16'), /* PU Adapter-Control-Space    */
   SP_ACS LITERALLY('17'), /* SP Adapter-Control-Space    */
   MBA_ACS LITERALLY('18'); /* MBA Adapter-Control-Space  */
DECLARE 1 ecs_event_vector    BIT(32), /* ECS Layout   */
   .1 bit_assignments,
      2 *                BIT(1),
      2 i370_s370_event  BIT(1),
      2 i370_timer_event BIT(1),
      2 any_programmed_event BIT(1),
      2 *                BIT(12),
      2 dummy_pu_request    BIT(1),
      2 sp_pu_request    BIT(1),
      2 mba_pu_request   BIT(1),
      2 *                BIT(13);
DECLARE Read_control_space ENTRY (INTEGER VALUE,
                   INTEGER VALUE)
                   RETURNS (INTEGER)
                   LINK ('A0223000'xc);
DECLARE Test_and_Leave_I370 ENTRY RETURNS (INTEGER)
                   LINK ('B2300020'xc);
DECLARE (SP_Request_Handler, MBA_Request_Handler,
   S370_Event_Handler, Timer_Event_Handler,
   PCS_Event_Handler, Dispatcher_Error_Routine) ENTRY;
ecs_event_vector = Read_control_space(ECS,0);
DO UNTIL (ecs_event_vector=0);
 SELECT;
  WHEN (sp_pu_request)      CALL SP_Request_Handler;
  WHEN (mba_pu_request)     CALL MBA_Request_Handler;
  WHEN (i370_s370_event)    CALL S370_Event_Handler;
  WHEN (i370_timer_event)   CALL Timer_Event_Handler;
  WHEN (any_programmed_event)   CALL PCS_Event_Handler;
  OTHERWISE CALL Dispatcher_Error_Routine;
 END SELECT;
 ecs_event_vector = Test_and_Leave_I370;
END DO UNTIL;
 END PROCEDURE Dispatcher;
```

Figure 203. I370 dispatcher

7.2.3.1 ES/9370 System Structure

Most of the various names of I370 control spaces mentioned in Figure 205 in the system's structure in Figure 206 ECS, TCS, and PCS are not shown; they are inherent to I370. All control spaces providing access to the system resources are explicitly shown:

- SCS as interface to S/370 hardware and microcode
- PU-ACS as interface to non-S/370 related CPU microcode
- SP-ACS as interface to the Service Processor (SP)

```
* Control Space Instruction

  - Read  Control Space (RCS)
  - Write Control Space (WCS)
  - And   Control Space (NCS)
  - Or    Control Space (OCS)
  - Xor   Control Space (XCS)

* Control Instructions

  - Leave I370 Mode (LIM)
  - Restore General Register Set (RGRS)
  - Save General Register Set (SGRS)
  - Load PSW from I370 assigned locs (LPIAL)
```

Figure 204. I370 unique instructions

- MBA-ACS as interface to the I/O controllers (IOPs) connected via a Memory Bus Adapter (MBA) chip, BCU chips, and I/O busses.

7.2.3.2 Service Processor to I/O Controller Communication

Let us assume a user interrogating the status of an attached I/O controller (IOP) with the help of the Service Processor (SP) console. The user selects "IOP Diagnostics" and enters a request (for example "QIYn" to display I/O controller status for channel n).

The following steps may be visualized by looking at the system's communication structure in Figure 207.

The Service Processor (SP) routes the inquiry via the SP-ACS to the CPU. Here, it causes an event indicating (in the ECS) that there is a request pending from the SP. The dispatcher calls the CPU/SP communication routine, which reads the SP-ACS to obtain further details. Finally, an I370 application, "SP/IOP communication", gets control and serves as a bridgehead in the CPU. It reroutes this request to the addressed I/O controller by assistance of the IOP/CPU communication routine. This service routine writes information into the MBA-ACS in order to set up a message that is sent via the I/O Bus to the subject I/O controller. When the I/O controller finally generates the answer, it sends back a message via the I/O Bus. The same kind of I370 programs in the CPU, which were used to transmit the inquiry from the SP to the I/O controller, will now make sure that the answer from the I/O controller will reach

```
* Architected Control Spaces

  - Event Control Space (ECS)
  - S/370 Control Space (SCS)
  - Timer Control Space (TCS)
  - Programmable Control Space (PCS)

* Implementation Specific Control Spaces

  - PU Adapter Control Space (PU-ACS)
  - SP Adapter Control Space (SP-ACS)
  - MBA Adapter Control Space (MBA-ACS)
```

Figure 205. I370 control spaces

the SP. All I370 control spaces mentioned above are again involved in transmitting the answer through the CPU to the SP.

7.2.3.3 Extending the Kernel Functions

All I370 code is written in PL.8, a high level language, best described as a PL/1 subset. An examination of Figure 207 shows that the CPU, including I370, provides various functional layers. The first and most important one has been mentioned earlier: the processor kernel consisting of CPU hardware and microcode. On top of that, two more layers extend the processor's functional capabilities. The lower one provides the base for all I370 application programs:

- PL.8 run-time environment (RTE),
- An event handler (dispatcher), which activates various I370 routines based on their associated events,
- Common I370 services - including message, timer, and trace facilities,
- Routines for IOP/CPU and CPU/SP communications.

The upper layer is the I370 application layer and contains all routines required to make it a full function system. It is split into two logical parts: "resident" and "temporary" I370 routines.

During IML, various functions have to performed which are not needed during system operation: (e.g. "I/O Bus tests" or "I/O controller initialization"). These routines are called "temporary" because they will be erased from main memory after IML is complete.

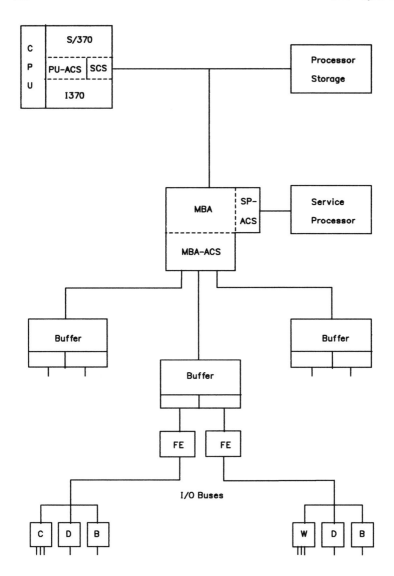

Figure 206. ES/9370 system structure

"Resident" routines are needed during normal system operation. These, as well as all I370 support/service routines, are kept permanently in processor storage.

Figure 208 is a complete list of I370 support and application programs:

High Level Microprogramming in I370

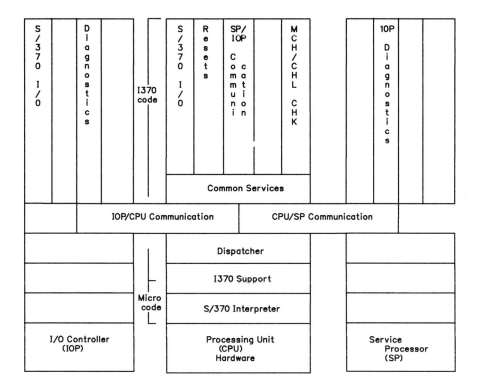

Figure 207. System communication structure

7.2.3.4 Simulation Concept for I370 Programs

One major objective during the initial design phase was to provide excellent simulation capabilities throughout the project. The I370 approach resulted in

- Execution in "real" S/370 processing speed on a S/370 host system (one simulation run requires a few CPU seconds instead of CPU minutes,
- Use of standard VM/SP services - only external, program, and I/O interrupt are intercepted,
- Availability of simulation support nine months prior to power on of the initial hardware.

Each I370 programmer is able to simulate "his" own system on a host computer independent of other users. All he needs is an ordinary terminal to control the I370 simulation session and an optional IBM PC/AT (or IBM PS/2), which acts as the ES/9370 system's Service Processor.

In order to simulate I/O operations with a virtual I/O controller, a logon on an additional VM/SP USERID is required. This virtual I/O controller operates as a finite state machine: the I/O controller's behaviour (including simulation of multiple devices attached to it) is described by a table stored on the programmer's disk space (A-disk). This table is interpreted by the simulator program

```
* Resident I370 Routines
  - I370 Support and Common Services
    - PL.8 Run Time Environment
    - Dispatcher
    - Timing Facility
    - Message Distribution Facility
    - Tracing Facility
    - Program Check and Manual Operations Handler
    - PU/SP Communication Routine
  - S/370 Support
    - I/O Instructions and Interruptions
    - Customer Manual Operations
    - Reset
    - Machine Check Handler
    - Diagnose (X'83') Instruction
    - PSW and I/O Trace
  - I/O Bus Support
    - PU/IOP Communication Routine
    - I/O Bus Error Recovery
    - IOP Manual Operations Support
    - I/O Diagnostic Monitor
    - IOP Load Support
    - IOP Re-IML
* Temporary I370 Routines
  - I370 Code (TOC) Loader
  - IML Monitor
  - I/O Bus Test Monitor
  - I/O Bus Test
  - IOP Initialization
```

Figure 208. I370 support and applications

executed in the simulated I/O controller. All S/370 debugging aids provided by VM/CMS are available during I370 simulation.

The complete simulation environment is outlined in Figure 209.

7.2.4 Conclusions and Outlook

When drawing conclusions on our experience with I370, we definitely have to look at:

- Productivity

    58 KLOC (lines of code) have been written in I370/PL.8, including 5 KLOC performance-critical I/O code. The productivity in terms of

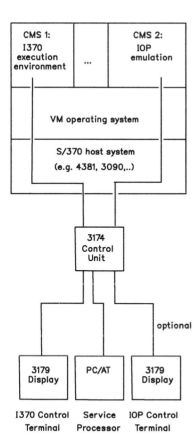

Figure 209. Simulation environment

KLOC/people/year including I370/PL.8 education, design, coding, and unit testing, is slightly higher (20-30%) than for assembler microcode. Thus, the overall productivity is much improved as one line of PL.8 code is equivalent to multiple lines of assembler code.

- Testability/Quality

  The "native" VM/SP simulation capability permitted to develop and unit test all major parts of the total I370 application package prior to power on. No functional problems in the first I370 code driver running on the initial hardware were encountered.

- Performance

  Figure 202 classifies 15% to be simple, performance critical code and 85% to be complex, performance non-critical code. The ES/9370 system's I/O code, however, does not fit into this scheme: it is complex and performance critical. We decided to implement this piece of code in I370 as well. Initial performance estimates yielded acceptable I/O degradation.

- Maintainability

    The remaining errors found during ES/9370 system tests could be fixed easily and reliably. Introduction of new functions during a late phase of the project could be managed efficiently.

- Portability

    We are confident that the I370 application programs written for the 9370 Processor are a viable base that will be maintained and further utilized in Boeblingen's follow-on S/370 systems.

In total, the I370/PL.8 concept allowed us to make a significant step forward in terms of productivity, quality, maintainability, and portability.

7.3 System Bring-Up and Test

Walter H. Hehl

7.3.1 Overview

Very little is written about the bring-up process in the computer engineering literature. After completion of design and simulation, the process of empirical analysis of system behaviour and problem resolution to create a running system is a complex technical and organizational task. Bring-up starts with the power-on of the chip set (or a subset) in a prepared bring-up station. Figure 210 shows a simplistic process flow from first silicon hardware availability to start of volume production.

Power-on is the starting point of the bring-up. In the hardware and microcode system, function by function is enabled and error after error eliminated. Basic bring-up is the elimination process of vital implementation errors or (if any) design errors. It produces a system running already some non-trivial test software, as, e.g., a stripped-down operating system. Design and bring-up groups continue the bring-up by performing a set of well-defined system test procedures, the engineering verification test (EVT). Successful completion of these tests terminates the bring-up process.

In the interest of test independence and objectiveness, technical test responsibility is transferred to a separate group of testers (Product Assurance), intentionally organized independent of the development laboratory and with a separate budget. From the beginning, this test group has been accompanying both product design and bring-up. They plan and perform a final quality test, the design verification test (DVT), executed on systems built by one or several manufacturing locations. As a rule, a rather large number of moderate or minor problems surface. Their elimination usually places an extremely heavy

System Bring-Up and Test 317

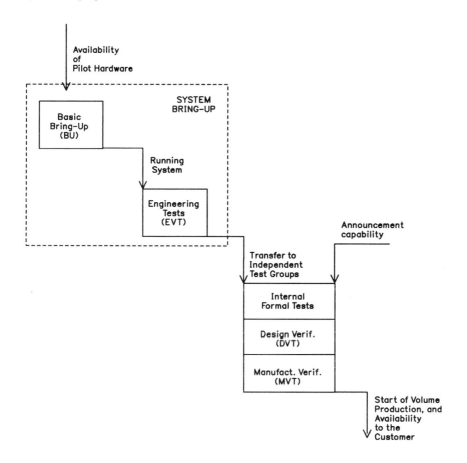

Figure 210. Development phases (schematic) from pilot hardware availability to general product availability

workload on the development organisation. The successful exit from DVT confirms the design maturity of the product and the achieved level of quality.

Product announcement to the market can take place based on intermediate reliability data from the DVT or, in general, after DVT completion. At least one of the manufacturing locations assures in parallel to the development and product quality tests the manufacturability of the product in the manufacturing verification test (MVT) for the start-up of manufacturing.

Successful MVT and DVT are the technical prerequisites for releasing the product to general manufacturing and general availability for the customer.

7.3.2 Bring-Up Strategy

Work on the CAPITOL chip set started in early 1983 by a small group of people. In mid 1983 a decision was made to implement a CMOS realisation (instead of the originally planned bipolar realisation). The process to obtain management approval for starting a major project began at the same time. A formal project organisation was formed in the fall, and staffing up to about 70 people was completed by the end of 1983.

We originally favoured a DCVS circuit tree, which in early 1984 was replaced by the existing approach. The final decision to use LSS and silicon compilation occurred in August 1984, and simulation, see "3.7 Logic Simulation" on page 190 started in early 1985. After producing several test site chips, first silicon of a full functional chip set was available in the fourth quarter of 1986. By April 1987 several systems were successfully executing a version of the VM/SP operating system.

A separate organisation to develop the IBM ES/9370 Model 30 system was started in mid 1986. Figure 211 shows the timing of the subsequent events.

The bring-up started with a basic Capitol chip set processor unit without floating point coprocessors and in a restricted system environment with limited I/O. Logic designers and the bring-up team debugged this first real hardware in close cooperation. Two system bring-up groups extended the system analysis to the first Capitol chip set-based products.

Test activities have correspondingly been divided into chip set type tests and into the engineering tests for the integrated systems. Organizationally, the system bring-up implied close cooperation between the Capitol chip set development group and the system development groups.

7.3.3 Basic Bring-Up Process

Power-on of the first real VLSI hardware from the pilot line is performed in a simplified hardware and microcode environment, the bring-up vehicle. It uses a modified ES/9370 processor cage supporting only a single IBM 9370 I/O bus, without the floating point coprocessor. The microcode is restricted to the most essential routines extracted from the microcode library. The I/O devices attached are only 2 types of devices, native 3270 terminals and artificial 9370 I/O test devices.

After debugging the individual cards and boards for wiring errors, the brand-new system is exposed to a hierarchy of verification levels with an increasing logical complexity of the performed operations (Figure 212). This hierarchy in testing is reflected in a corresponding hierarchy of test tools (test devices and test programs).

System Bring-Up and Test

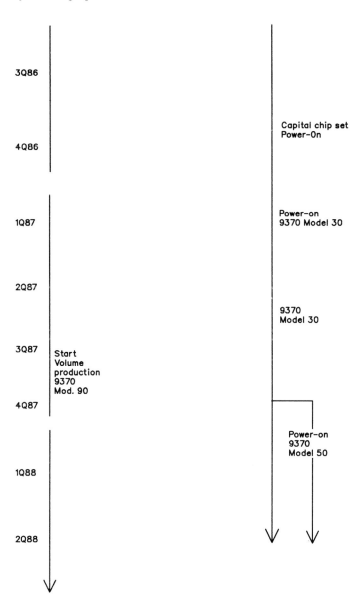

Figure 211. Bring-up schedule

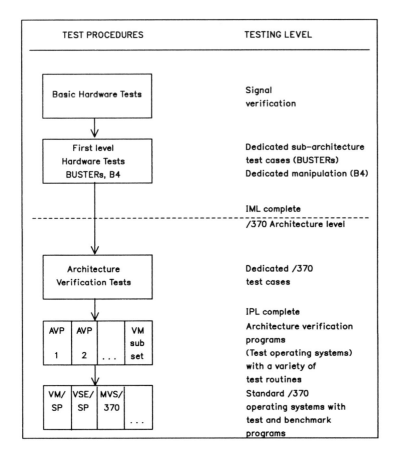

Figure 212. Bring-up and system verification levels

7.3.3.1 Sub-Architectural Verification

Capitol chip set unique first level hardware "test cases" (sets of microcode or machine instructions) have been prepared by the logic designers to stress precisely those sub-architectural hardware elements in the processor unit participating in a test. Most of these BUSTER test cases (Bring-Up Supporting Tests and Exerciser Routines) have already been heavily used for the simulations (see "3.6.5 What Drives the Simulation - Testcases" on page 188) and for characterization (see "5.2.1.4 Functional Pattern Test" on page 258). The test cases are by definition self-driving and autonomous, i.e., after load in the micro store from a PC they are executed without further prerequisite and without a status change of the system after execution. For the convenience of the operation in bring-up, and, in exceptional cases, also in manufacturing, a test case monitor allows definition of arbitrary test case sequences and then running them once or an arbitrary number of times.

System Bring-Up and Test 321

For test result analysis and manual interventions, the processor internal data (e.g. register and storage contents) can be displayed and altered through a special support program.

7.3.3.2 Architecture Verification

With the set of about 500 test cases correctly executed by the real hardware, the bring-up checkpoint "All BUSTERs COMPLETE" is achieved. This completes the pure hardware debug phase and the processor system is ready to run the microcode. The initial microcode load (IML) begins with the initialization of the processor itself and completes with the enabling of the attached I/O. After successful debug of the IML step and initialization of the functional microcode, the system is ready to execute the set of S/370 instructions and start architecture verification through specific architectural test cases. All new S/370 processor designs are verified with these testcases.

In practice, a processor with verified hardware through successful test case runs executes the standard simulated VESTAL test cases right away and allows the Initial Program Load (IPL) of simple test operating systems.

7.3.3.3 Testing Under the PAS Control Program

Functional verification has been mainly accomplished using the Product Assurance System (PAS) control program. Developed by the S/370 laboratories Boeblingen, Endicott, and Poughkeepsie, PAS runs functional tests for architecture verification, I/O functional tests and data pushers to test the I/O functions and to create I/O load, and I/O fast data pushers to create heavy I/O load. PAS is a dynamic test package creating its own flow of test instructions and their expected results, as compared to static tests prepared off-line (as, e.g., VESTAL tests).

PAS tests have already been performed during system simulation ("3.7 Logic Simulation" on page 190) using the powerful test package PGEN with random instruction streams and random data parameters. Because of computing time restraints, this simulation was equivalent to about 40 seconds (500 million clockcycles) real CPU time. On the actual hardware, the complete spectrum of test exercisers available could be applied for almost infinite time, i.e., several thousand hours of system run time.

In order to evoke system stress situations, selected sets of processor and I/O exercisers are run jointly. Exercisers contained in the stress scenario include:

- General processor instruction streams for functional verification

- Allowed, but unusual or even meaningless instruction sequences (garbage streams)

- Storage and storage protection tests to stress the Processor Bus and storage control

- Extensive S/370 channel tests (system side) driving the system channels with special channel test devices
- Extensive ES/9370 I/O bus tests, driving the I/O busses with special bus test devices.

7.3.3.4 System I/O and Interaction Testing

Dedicated testing of the operations of the new system in conjunction with all specified I/O control units and I/O devices is performed by static or dynamic exercisers comparing observed and predicted channel, I/O, storage, and PSW (Program Status Word) data. Device specific I/O test packages stress the system up to artificial overload conditions used to determine inherent system limits.

Standard operating systems and standard workloads under these operating systems that use the system resources in a practical way serve as a supplement to artificial exercisers. In particular, the VM/SP operating system has been used in basic bring-up to functionally test in particular paging and cache mechanisms.

7.3.4 System Bring-Up

After all major problems are either solved or circumvented by hardware or microcode fixes, bring-up continued with the Engineering Tests (EVT) on the base of the first systems using the Capitol chip set, especially the IBM ES/9370 model 30.

The functional extension from the basic bring-up to systems bring-up included adding full floating point hardware support through dual floating point unit chips, up to 4 native I/O busses, generic ES/9370 I/O including LANs and TP, generic S/370 operating systems with appropriate microcode assists, and system maintenance support verified through injected errors. Some of these new functions required additional bring-up, all of these added new microcode to the system and all needed further testing.

System machine check and maintenance tests are performed by introducing carefully selected hardware and microcode errors into the system, e.g. via "bugged" processor cards.

The total set of testing activities separated naturally into two classes, chip set tests (as, e.g., the clock variation analysis, see "2.9.7 Clock Variation" on page 125), and system tests (as, e.g., electromagnetic compatibility of processor cards). A carefully planned network of tests was established to guarantee both chip set and system quality. Large system configurations employing many I/O devices on all channels are tested with multiple operating systems and under realistic workloads. This completes the bring-up process, demonstrating the correctness of the design under complex system interaction situations.

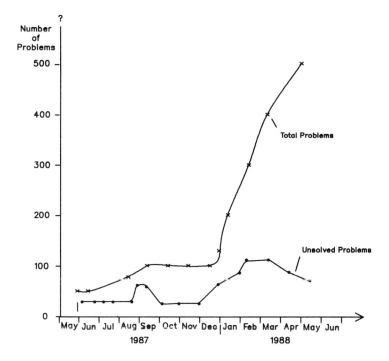

Figure 213. Problem statistics evolution in IBM 9373 Model 30 bring-up time

Next, as indicated in "7.3.1 Overview" on page 316, test responsibility is turned over to the Product Assurance organisation. They perform a final quality test (several month duration) using machines built by a manufacturing location.

7.3.5 Regression Testing

System bring-up represents the first of many periods of thorough system testing. Design changes are either introduced to remove design errors or enhance the design. Unfortunately, every design change that is made to correct a deficiency has a non-zero probability to generate a new one. Complementing the waves of design changes and functional system enhancements, a sequence of rigorous test waves is required. After each change or enhancement, a regression test has to be performed for the function changed itself, but also to verify that all other functions are not affected. These iterations of test procedures require automation. The Automatic Test System (ATS) simulates the various operations that can be started from the ES/9370 system console. These operations range from standard manual operations to functions or test cases only accessible for customer or development engineers, including special test cases.

MAJOR EC LEVEL (Silicon change)	0				1		2
MINOR EC LEVEL (Metal layer change)	0	1	2	3	0	1	0
CPU Chip	X	X	X	X			
MMU Chip	X	X	X	X	X		
STC Chip	X	X					
Clock Chip	X				X		X
FPU Chip	X				X	X	
MBA 1 Chip	X				X		X
MBA 2 Chip	X				X		
MBA 3 Chip	X				X		
MBA 4 Chip	X				X		

Figure 214. Engineering levels of the chips of the Capitol chip set

The system responses (e.g. total screens, system messages, reference codes) are compared with stored predicted results. Test cases and results are logically attached to the system micro code library and promoted in parallel to the system releases. The set of test programs applied during an automatic test shift of 2 hours can be equivalent to a week of manual testing.

7.3.6 Bring-Up Results and Error Corrections

Several months after the start of the basic bring-up, a formalized procedure for reporting design problems was introduced. The evolution of problem statistics as a function of bring-up time reflects the major phases of the process (Figure 213). The number of total problems is divided into hardware problems and microcode problems. At the start of the system bring-up (design verification test), all open problems were of the microcode type.

7.3.6.1 Basic Bring-Up

The initial bring-up period has been dominated by the hardware: the problems are almost exclusively of this type. Early Engineering Test activities lead to a first peak in the unresolved problem graph. Only a small fraction, namely 5 hardware problems turned out to be logic errors affecting the VLSI structures. Engineering change levels of the individual VLSI chips of the Capitol chip set processor are shown in Figure 214.

The majority of the corrections could be implemented by rerouting the wiring on the first and second metal layers; only some changes required modifications of the underlying silicon. A small fraction of those had been stimulated by the bring-up groups. Others, especially at the later levels, were introduced as performance enhancements to stabilize the cycle time across the complete chip set. Corrections to some errors detected at later bring-up periods could conveniently be incorporated into these final higher performance chip levels for a final clean-up.

7.3.6.2 System Bring-Up

The second peak in the graph of the number of unresolved problems (Figure 213) marks the beginning of the Design Verification Test (DVT) activities. At this point in time, the hardware system had achieved a mature status with the first early manufactured systems delivered from the plant for testing purposes. The number of microcode problems continued to rise sharply caused by the inclusion of the full spectrum of system functions from system configuration to remote services. New problems have been almost exclusively microcode errors, most of them in non-critical areas invisible to the user under regular conditions. A large fraction of these problems refers to the correct analysis and treatment of system error situations for maintenance purposes.

The debug process for errors in the microcode system consists of identifying the defective program module (and thereby the owning code designer), error correction, and solution verification. Microcode debug misses the dramatics of the hardware error analysis where a hard problem might jeopardize the schedule of the project by an unscheduled VLSI engineering change. The major challenge in microcode bring-up and testing is the administration and simultaneous handling of a large number (e.g. more than hundred) of active problems in very different functional areas owned by a variety of organizations.

7.3.7 Summary and Conclusions

Bring-Up has been implicitly defined as the planned process to resolve unplanned technical problems.

With respect to the development of the processors of the next generation, the Capitol chip set bring-up demonstrated (again) the importance of thorough simulations and pointed to the critical functional hardware areas. Availability of state-of-the-art tools is mandatory for simulation and bring-up.

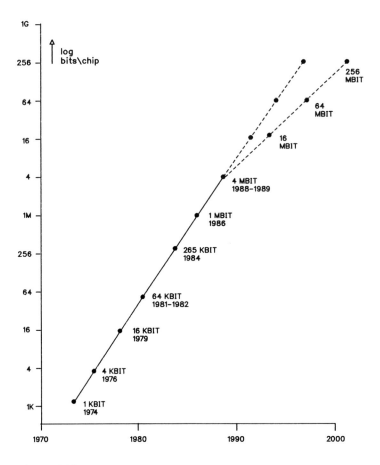

Figure 215. Memory chip density

For the overall system, in particular the microcode structure, the rigorous top-to-bottom functional design and the automatic control of the multitude of elements proved to be vital and mandatory.

7.4 Outlook

It is always difficult to predict the future. However, there is broad consensus that the rapid technological progress in silicon technology will continue at more or less the same rate, at least for another decade. The progress of semiconductor memory technology, shown in Figure 215, represents a steady annual compound growth rate of more than 70% during the last 20 years.

The growth in CPU speed for S/370 machines is shown in Figure 216. (Please note that performance comparisons of machines with different architectures are extremely difficult.) The performance of high end uniprocessors has been

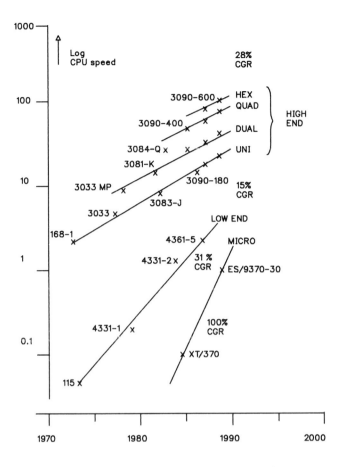

Figure 216. S/370 processor performance growth

increasing at a compound rate of 15% annually. The introduction of 2way, 4way and 6way tightly coupled parallel processors has provided a 28% annual performance growth for high end systems (as opposed to high end uniprocessors).

Low end uniprocessors (minis or superminis) have exhibited a much higher performance increase of 31% per year. This however is still small compared to the approximately 100% performance growth rate of microprocessors.

Two factors have contributed to this growth. Shrinking semiconductor dimensions lead to faster devices and shorter machine cycle times. In addition, increasing device densities and larger chip sizes permit a significant increase in CPU (and MMU) circuit count, and thus CPU performance.

The 100% annual microprocessor performance increase is a phenomenon generally observed within the industry. The coming 16 Mbit/chip technology will result in maybe a factor 10 area density advantage over todays 1 Mbit/chip technology. A factor 10 will permit to put existing high end system structures (e.g. IBM 3090) on a single chip. 64 and 256 Mbit/chip technologies are being discussed. Using these advanced technologies in future microprocessors will also require considerable extensions and improvements in the design approaches and tools described in this book. At the same time it will become a mundane and relatively inexpensive development task to design chips with the complexity of the Capitol chip set. It will be exciting to participate in the opportunities, which these future technological capabilities will create for both microprocessors and for other VLSI chip structures.

Authors

Günter BILGER
received his Dipl.-Ing. degree on electrical engineering from the Stuttgart University and joined IBM in 1977. He is presently a member of the Boeblingen Lab VLSI Logic Chip Development department.

Dietrich BOCK
joined IBM in 1968 having received his Dipl.-Ing. degree in electrical engineering from the Stuttgart University. He received a master's degree in electrical engineering from the University of Vermont, USA, in 1980. Presently he is a staff member of the Boeblingen Lab Processor Development department.

Michael FAIX
joined IBM in 1962 and worked in the areas of OS/360, computer architecture and evaluation, /370 processor and system design, and system RAS. He is presently a Senior Technical Staff Member of the Boeblingen Lab Processor Development department.

Horst FUHRMANN
born 1944, he received the Dipl.-Ing. degree of Electrical Engineering from the University of Karlsruhe in 1969 and joined the IBM Laboratories in 1970. After several tasks in memory subsystem design, he joined a design group in 1983, whose studies led to the development of the Capitol Chip Set.

Bernd GARBEN
received his Dipl.-Phys. degree and Ph. D. degree in physics from the University of Goettingen, and joined IBM in 1974. He is presently a staff member of the Boeblingen Lab Semiconductor Technology Development department.

Walter HEHL
received his Ph. D. in physics from the Stuttgart University, and joined IBM in 1973. After working as a development engineer on printer technology and digital image processing, he became responsible for the bring-up of the systems IBM 9373 model 30 and IBM 9375 model 50 using the VLSI /370 chip set. He is presently manager of a Boeblingen Lab department for new ES/9370 system applications.

Siegfried HEINKELE
received his Dipl.-Inform. degree in Informatik at the Stuttgart University in 1981, and joined the IBM Laboratories the same year. After working for two years in Product Assurance, he moved to the Hardware Design Tools department, supporting processor development with front end tools to the IBM Engineering Design System (EDS).

Klaus HELWIG
received his Dipl.-Ing. degree in electrical engineering from the Technical University of Vienna and joined IBM in 1968. He is presently a development engineer of the Boeblingen Lab, VLSI Array Design department.

Michael J. KESSLER
received his Dipl.-Ing. (FH) degree in 1970 and joined IBM in the same year. He first was working in Equipment Engineering for short-open testers for printed circuit boards. Later he served as a specialist for test data generation and design for testability. Presently he is manager of the Test Methodology and Test Generation department.

Bernhard KICK
received his Dipl.-Ing. degree in Electrical Engineering from the Technical University of Munich in 1983. In 1985 he joined IBM, where de develops and maintains software tools for digital logic hardware design.

Klaus KLEIN
joined the IBM Boeblingen Laboratory in 1964. Since 1965 he has been involved in computer-aided design systems applications, mainly in the area of physical design and component packaging. Since 1979 he has been in charge of placement and wiring studies, and later on physical design methodology development for VLSI chips.

Gunther KOETZLE
received his Dipl.-Ing. degree in electrical engineering from the Stuttgart University, and joined IBM in 1968. He is presently manager of the Boeblingen Lab VLSI Logic Chip Development department.

Hans KRIESE
received his Dipl.-Phys. degree from the Tuebingen University, and joined IBM in 1973. He has worked on design and verification concepts, and is presently a staff member of the Boeblingen Lab Processor Development department.

Karl E. KROELL
received his Ph. D. degree from the University of Marburg, and joined IBM in 1968. He worked on hot processes in the Boeblingen pilot line, on semiconductor device design and electrical failure analysis. Presently he is a staff member of the Boeblingen VLSI Device and Array Design department.

Wolfgang KUMPF
joined IBM in 1970 as a customer engineer, worked as a technician in the Boeblingen Lab Printer Development and Memory Development. He left IBM for a university education in electrical engineering and received the Dipl.-Ing. degree from the Stuttgart University. He joined IBM again in 1980 and is presently working in the Boeblingen Lab Processor Development department.

Heinrich LINDNER
received his Dipl.-Ing. and Dr.-Ing. degrees in physics from the Technical University of Berlin, and joined IBM in 1967. He is presently manager of a group for high density CPU design.

Thomas LUWDIG
received his Dip.-Ing. degree in electrical engineering from the Technical University of Berlin in 1984 and joined IBM in the same year. He is presently a staff member of the Boeblingen Lab VLSI Logic Chip Development department.

Jürgen MÄRGNER
is a senior engineer at the Boeblingen Lab, where he is currently working on channel subsystem design for S/370 based systems. He received his Masters of Science degree in electrical engineering from the Technical University of Berlin and joined IBM in 1970.

Klaus-Dieter MÜLLER
received his Dipl.-Ing. degree in electrical engineering from the Stuttgart University, and joined IBM in 1981. He is presently a manager in the Boeblingen Lab Component Development department.

Rolf MÜLLER
received his Dipl.-Ing. degree in electrical engineering from the University of Hannover. He joined IBM in 1978 and worked for several years on dynamic memory design, testing, and logic design. Presently he is working in the computer science and communication group in IBM Research Rueschlikon.

Jean-Luc PETER
received his Engineer degree from the Ecole Superieure d'Electricité (ESE) in 1967 and a Business Administration degree from the Faculté de Droit et de Sciences Economiques from Paris in 1969.
He joined the IBM Component Development Lab in Corbeil Essonnes in 1969, where he was involved in the design of SRAM's and later on, in the design of the floating point chips for the ES/9370 systems.
In 1988 he joined the Communication Product Division in La Gaude, where he is now working on hardware development of Communication Controllers.

Kurt POLLMANN
received his Dipl.-Ing. degree in electrical engineering from the Technical University of Berlin, and joined IBM in 1978. He is presently a staff member of the Boeblingen Lab VLSI Logic Chip Development department, working on hierarchical layout and checking of ASIC CMOS designs.

Joachim RIEGLER
received his Dipl.-Ing. (BA) degree in electrical engineering from the Staatliche Studien-Akademie Stuttgart and joined IBM in 1981. Since 1984 he has been serving as a specialist in test development and VLSI logic characterization in the Boeblingen Lab Component Development department.

Wolfgang RÖSNER
received his Dipl.-Ing. degree in electrical engineering from the University of Kaiserslautern in 1980. In 1983 he received the Dr.-Ing. degree in electrical engineering from the University of Kaiserslautern.
In 1984 he joined the IBM Laboratories in Boeblingen where he is is presently a staff member of the Processor Development department.

Peter H. ROTH
received his Dipl.-Ing. degree in electrical engineering and his Dr.-Ing. degree from the Stuttgart University in 1979 and 1985 respectively.
In 1985 he joined the IBM Boeblingen Lab, starting in the department of VLSI logic chip development. Since 1987 he has been leading the VLSI test and characterization team of the Boeblingen Lab. Dr. Roth is a member of the Institute of Electrical and Electronics Engineers.

Peter RUDOLPH
joined IBM in 1961 as a customer engineer. He served as a specialist on reliability, availability, and serviceability for the development of low end S/360 and S/370 systems. He is presently a staff member of the Boeblingen Lab Processor Development department.

Helmut SCHETTLER
received his Dipl.-Ing. degree in electrical engineering from the Stuttgart University, and joined IBM in 1969. He is presently a staff member of the Boeblingen Lab VLSI Logic Chip Development department.

Dietmar SCHMUNKAMP
received his Dipl.-Ing. degree in electrical engineering from the Darmstadt University, and joined IBM in 1984. He is presently a development engineer of the Boeblingen Lab Component Development department.

Uwe SCHULZ
studied electronics in El Paso, Texas, USA, and electrical engineering in Darmstadt. He joined IBM in 1963 and is presently engaged in VLSI CAD tools development as a member of the VLSI Logic Chip Development department.

Hermann SCHULZE-SCHOELLING
joined IBM in 1967 as a field engineer and came to the Boeblingen Lab in 1971. He served as a specialist on microcode and hardware design for the development of low end /370 systems. For 10 years he had been involved in the design of a single chip CPU on a large chip. He is presently a staff member of the Boeblingen Lab Processor Development department.

Michael SCHWARTZ
obtained his Ph. D. in Materials Science from Rensselaer Polytechnic Institute and joined IBM in 1968. He worked in several development programs throughout IBM, and managed the development team, which qualified the CMOS technologies in IBM. Currently, he is a project manager in Advanced Memory Development in IBM Essex Junction, Vermont, USA.

Hartmut SCHWERMER
Ph. D., is a senior engineer at the Boeblingen Lab, where he is currently working on system design for S/370 based systems. He received his Masters degree in mathematics from the Technical University of Berlin and joined IBM in 1970.

Claude SITBON
received his Dipl.-Ing. degree in electrical engineering from the Ecole Superieure d'Electricite Paris, and joined IBM in 1978.
He served as a specialist on Microprocessor design. He is presently an engineer of the Corbeil Essonnes Component Development Lab.

Wilhelm G. SPRUTH
received his Dr.-Ing. degree in electrical engineering from the Aachen Technical University and joined IBM in 1959.
He was the Capitol chip set project manager during the first part of its development phase. Presently, he is manager of Advanced Technology at the Boeblingen Lab.

Rainer STAHL
received his Dipl.-Ing. degree in electrical engineering from Stuttgart University, and joined IBM in 1974. He worked on packaging aspects of low end S/370 systems. In 1984 he became manager of a department developing main memory cards. In 1986 he became manager of VLSI Packaging in the Boeblingen Lab Processor Technology department.

Donald THOMAS
obtained his MSEE from the Colorado State University in 1968, and joined IBM in 1970. He has managed various device physics, device modelling groups most recently in CMOS technology. Currently he is a senior engineer / manager of the CMOS Logic Device Physics Group in IBM Essex Junction, Vermont, USA.

Otto WAGNER
received his Dipl.-Ing. degree from the Technical University Munich in 1969. In 1969 he joined the IBM Sindelfingen Plant, working in the field of test and characterization of bipolar and FET memory chips. In 1978 he joined the IBM Boeblingen Lab, where he is engaged in CMOS VLSI circuit and chip development as a staff member in the VLSI Logic Chip Development department.

Dieter F. WENDEL
joined IBM in 1981 as a design engineer in the Fellowship VLSI department. He was on assignment to the IBM Research Center in Yorktown Heights from 1984 to 1986. Since his return he serves as a staff member in the Boeblingen Lab Test and Characterization department.

Rainer ZÜHLKE
received his Dr.-Ing. degree in electrical engineering from the Stuttgart University. He joined IBM in 1970 and is presently working in the Boeblingen Lab VLSI Logic Chip Development department.

References

[AMDA] G.M. Amdahl, G.A. Blaauw, F.P. Brooks: "Architecture of the IBM System/360". IBM J. Res. Develop. 8 (2), 87-101 (1964).

[BAER] J.L. Baer: "Computer Systems Architecture". Computer Science Press (1980).

[BELL] C. Bell, J. Craig Mudge, John E. McNamara: "Computer Engineering". Digital Press, (1978).

[BLAA] G.A. Blaauw: "Computer Architecture". In Hasselmeier, Spruth (Ed.), Rechnerstrukturen, R. Oldenbourg (1974).

[BLAA] G.A. Blaauw: "Digital System Implementation". Prentice-Hall (1976).

[BRAN] D. Brand: "Redundancy and don't cares in Logic Synthesis". IEEE Trans. Comp. C-32 (10), 947-952 (1983).

[BREU] M. Breuer (Ed.): "Design Automation of Digital Systems". Prentice Hall (1972).

[CASE] R.P. Case, A. Padegs: "Architecture of the IBM System /370". Comm. ACM 21(1), 73-96 (1978).

[COWL] M.F. Cowlishaw: "The design of the REXX language". IBM Systems Journal 23(4), 326-335 (1984).

[DAR1] J.A. Darringer, W.H. Joyner, C.L. Berman, L. Trevillyan: "Logic Synthesis through Local Transformations". IBM J. Res. Develop. 25(7), 272-280 (July 1981).

[DAR2] J.A. Darringer, D. Brand, W.H. Joyner, J.V. Gerbi, L. Trevillyan: "LSS: A System for Production Logic Synthesis". IBM J. Res. Develop. 28(5), 537-545 (1984).

[DONZ] R.L. Donze, G. Sporzynsky: "Master-image Approach to VLSI Design". IEEE Computer 16(12), 18-25 (1983).

[GIFF] D. Gifford, A. Spector: "Case Study: IBM's System/360-370 Architecture". Commun. ACM 30(4), 292-307 (1964).

[HÖRB] E. Hörbst, C. Müller-Schloer, H. Schwärtzel: "Design of VLSI Circuits". Springer (1978).

[IBM1] "IBM System /370 Principles of Operation". IBM Corporation Form No. GA22-7000.

[IBM2] "IBM System /370 RPQ High Accuracy Arithmetic". IBM Corporation Form No. SA22-7093.

[IBM3] "IBM Mathematical Assist, Section Square Root". IBM Corporation Form No. SA22-7094.

[IBM4] "IBM Corporation VM/SP System Product Interpreter User's Guide". IBM Corporation Form No. SC24-5238.

[IEEE] "VHDL: The VHSIC Hardware Description Language". IEEE Design & Test (April 1986).

[JOIN] W.H. Joyner, W.H. Trevillyan, D. Brand, T.A. Nix, S.C. Gundersen: "Technology Adaptation in Logic Synthesis". Proceedings of the 23rd Design Autom. Conf., 94-100 (June 1986).

[MAIS] L.I. Maissel, H. Ofek: "Hardware design and description languages in IBM". IBM J. Res. Develop. 28(5), 557-563 (1984).

[MEAD] C. Mead, L. Conway: "Introduction to VLSI Systems". Addison-Wesley (1980).

[RUSS] G. Russel (Ed): "CAD for VLSI". Van Nostrand Reinhold (1985).

[SCHU] C. Schuenemann: "Micro- und Pico-Programmspeicher". In Hasselmeier, Spruth (Ed), Rechnerstrukturen, R. Oldenbourg (1974).

[SPIRO] H. Spiro: "Simulation integrierter Schaltungen". R. Oldenbourg (1985).

[TANE] A.S. Tanenbaum: "Structured Computer Organisation". Prentice-Hall (1984).

[TREV] L. Trevillyan, W.H. Joyner, L. Berman: "Global Flow Analysis in Automatic Logic Design". IEEE Trans. Comp. C-35(1), 77-80 (1986).

[WEBR] H. Weber: "A Tool for Computer Design". Proceedings of the 11th Design Automation Workshop, Boulder Co., 17-19 (June 1974).

[WEND] S. Wendt: "A New Simulation System for Hierarchical Models of Discrete Systems". Proceedings of the SCSC, Seattle, Wa. (1980).

Glossary

AC. alternating current

ACB. address check boundary

ACRITH. high-accuracy arithmetic

ADDR. address

AIX/370. advanced interactive executive / System 370

ALU. arithmetic logic unit

AS/400. application system/400

ASIC. application specific integrated circuit

ATS. automatic test system

BA. bus adapter

BBA. bus to bus adapter

BC. best case

BCU. bus control unit

BDL/S. basic design language for structure

BSM. basic storage module

BSR. basic storage register

BU. bus unit

BUSTER. bring-up supporting test and exerciser routine

CAD. computer aided/assisted design

CAS. column access strobe

CAW. channel address word

CC. condition code

CCW. channel command word

CGR. compound growth rate

CISC. complex instruction set computer

CMD. command

CMOS. complementary metal oxide semiconductor

COC. checking of checkers

CPU. central processing unit

CS. control store

CSAR. control store address register

CSW. channel status word

CTL. control

CTLR. controller

DASD. direct access storage device

DAT. dynamic address translation

DC. direct current

DCVS. differential cascade voltage switch

DED. double error detection

DFT. design for testability

DI. data in

DISCO. digital slope control

DL. data latch

DLS. data local store

DMA. direct memory access

DO. data out

DOG. data out gate

DPPX/370. distributed processing programming executive/System 370

DVT. design verification test

E-phase. execution phase

ECC. error checking & correction

ECS. event control space

EDFI. error detection and fault/FRU isolation

EDS. engineering design system

EDX. energy dispersion X-ray

EPROM. erasable programmable read only memory

ES/9370. enterprise system / 9370

ESD. electrostatic discharge

EVT. engineering verification test

EXP. exponent

FET. field effect transistor

FOP. forced operation

FPU. floating point unit

FRU. field replaceable unit

GL/1. graphics language/1

GND. ground

Hz. Hertz

HLL. high level language

HZ. high impedance

I-phase. instruction phase

I/F. interface

I/O. input output

IAR. instruction address register

IML. initial microprogram load

IOP. input output processor

IPG. interactive presentation graphics

IPL. initial program load

I370. internal 370

K. kilo (1,024)

KByte. kilo byte

KLOC. thousand lines of code

LAN. local area network

LRU. least recently used

LSAR. local store address register

LSS. logic synthesis system

LSSD. level sensitive scan design

LT. latch

M. mega (1,048,576)

MByte. mega byte

MBA. memory bus adapter

MC. metalized ceramic

MFLOPS. million floating point operations per second

MIPS. million instructions per second

MMIO. memory mapped input output

MMU. memory management unit

MOD. modifier

MUX. multiplexer

MVS/370. multiple virtual storage / System 370

MVT. manufacturing verification test

PAS. product assurance system

PC. personal computer

PCS. programmable control space

PER. program event recording

PGA. pin grid array

PGEN. basic processor test

PI. primary input

PL/1. programming language 1

PLA. page look ahead

PLA. programmable logic array

PO. primary output

PS/2. personal system 2

PSW. program status word

R/W. read/write

RAM. random access memory

RAS. reliability, availability, and serviceability

RAS. row access strobe

REG. register

REXX. restructured extended executor language

RISC. reduced instruction set computer

RIT. release interface tape

ROS. read only storage

RPT. random pattern testing

S/370. system/370

SAR. storage address register

SB. support bus

SBA. support bus adapter

SCS. S/370 control space

SEC. single error correction

SEM. scanning electron microscopy

SERDES. serializer/deserializer

SMI. system measurement interface

SP. service processor

SPARC. scalable processor architecture

SQRT. square root

SRL. shift register latch

STC. storage controller

SU. shift unit

TCS. timer control space

TEM. transmission electron microscopy

TLB. translation look-aside buffer

TOD. time of day

TP. teleprocessing

TSD. tri-state driver

TTL. transistor transistor logic

USI. unit support interface

V. Volt

VAX. virtual address extension

VCSEM. voltage contrast scanning electron microscopy

VESTAL. verify E-system to architecture level

VHDL. VHSIC (logic) hardware description language

VHSIC. very high speed integrated circuit

VLSI. very large scale integration

VM/CMS. virtual machine facility/conversational monitor system

VM/SP. virtual machine/system product

VSE/SP. virtual storage extended/system package

W. watt

WC. worst case

Index

AC test   245, 257
ACRITH   15, 99
action   191
address check boundary register   59
address compare register   58
address fault protection   95
address translation   50, 60
addressing stage   29
AND/OR level   161
any check   135, 140
architectural testcases   321
architecture   7
architecture verification   321
array   140, 155, 206, 215, 292
array integration   221, 247
array macro   63, 203, 247, 293
ASIC   281
assertions   149
attention interrupt   140
attributes   151
auto diagnostic   255

B

B-reg bus   20
banks   81
base level   26, 33
BCU chip   108, 301
BDL/S   145, 158, 171, 186, 190, 193, 281, 299
behavioral information   186
behavioral language   145, 156, 190, 192
block   190
books   292

boolean minimization   171
Booth encoding algorithm   105
boxes   158
bring-up   299, 316, 318, 325
bus unit   39
bus-to-bus adapter BBA   108

C

C-4   227
cache   12, 17, 53, 209, 215, 292, 302
cache array   70
cache directory   54, 60, 67, 215, 292
CAD   201, 204, 209, 281, 283, 299
capacitance   233
capacitor   211, 233
card enclosure   301
cell   170, 173, 205, 207, 293
central data bus   20
ceramic substrates   229
change bit   64, 67
channel command word   109
channel I/O   302
characteristic   100
check coverage   139
check restart   135
checker   135, 139
chip assembly   291
chip image   204, 292, 299
chip layout   292
chip pad   205, 207
chip partitioning   149
circuit analysis program   295
circuit model   295
clock chip   14, 112
clock comparator   44
clock distribution   122

clock macro   113
clock signals   118
clock skew   120
clock variation   125, 322
CMOS process   264
column access strobe   82
command   35
common invalid bits   65
compare logic   60, 66, 67
compartment   54, 70, 218
complement retry   92
cone of influence   248, 262
control microinstruction   35, 48, 65
control spaces   304, 305
control store   8, 14, 21, 25, 137
control store address register   27, 30
control store buffer   25
control store bus   9, 16
control unit   301, 322
CPU chip   10, 19
CPU timer   44
critical path   173, 263, 294
CS chip   14
CSAR   29

D

data integrity   85
data latch   82
data local store   10, 20, 31, 134, 137, 206, 210
data out gate   82
dataflow   170, 176
DCVS circuit   318
debugging   184
decode/select level   161
decoupling   233
decoupling capacitor   211
delay   173, 176
delay calculator   177, 294
delay equations   177
delay line   18, 112
depopulation   206
design language   145, 148, 156, 158, 171, 187, 193
design language compiler   155
design level   148, 149, 155

design system   171
design verification test   316, 325
detection   112
device properties   268
digital slope control   214
direct memory access   108
distance of a code   88
distributed simulation   199
double errors   92
double word crossing register   71
dynamic address translation bit   64, 66, 73

E

EDS   281, 283
electrostatic discharge protection   213
engineering change capability   299
engineering verification test   316
EPROM   21, 25, 114, 137
error correction   79, 88, 139
error detection   139
error recovery   112, 139
event control space   306
event pending   306
exception   26, 135
execution phase   21
exponent   100
exponent dataflow   104

F

failing location   262
failure analysis   270, 271
failure localisation   259
failure localization   271
failure mechanism   270
failure models   244
fanout   179
fanout correction   162, 166, 168
fault simulation   245, 247
fetch aligner   72
fetch protection bit   64
first silicon   318
floating point format   98
floating point registers   31, 43
floor space   171, 176

floorplan   179
footprint   229
forced operation   21, 26
FPU chip   15, 43, 99
fraction   100
functional level   186
functional patterns   245, 257, 258

G

gate array   204
general registers   31, 305
global nets   168, 179, 281
global wiring   300
groundrules   293
guard digit   104

H

Hamming code   91, 139
hard errors   85
hardware mode   21
high impedance   118
high impedance state   128
high level simulation   186
HP Precision   51
Huffmann sequential network   246

I

IBM 6150   51
IEEE 754   98
image module   292
immediate data bus   20
implantation   285
implementation   7
initial microprogram load   11, 96, 134, 301, 304, 311
instruction phase   21
Intel   51, 112
interconnect-areas   285
interconnect-pins   285
interleaving   13, 84
internal probing station   261
interval timer   44
inverted page table   51
IPL   135, 304
I370   11, 12, 24, 109, 303

I/O adapter   108
I/O busses   301, 322
I/O cell   206, 213
I/O controller   301, 310
I/O device   322
I/O driver circuit   211
I/O redistribution   205, 249

K

key store   12, 54, 73, 215, 292

L

latch-up   210, 264, 268, 278
late select   67, 218
layout rules   267
least recently used   12
leave I370 mode   308
load reset   135
local nets   179
logic book   205, 207, 262, 263, 292
logic design   145
logic design verification   145, 175, 185
logic gate   190, 204, 207, 247
logic simulation   145, 185, 222
logic synthesis   145, 149, 157, 158, 183
logout area   140
low level simulation   186
LRU   12, 51, 53, 68
LSS   148, 151, 155, 158, 171, 318
LSSD   18, 113, 135, 151, 218, 243, 246, 258, 263, 299
LSSD rules   247

M

machine check   140, 304
machine cycle   19, 112, 119, 183, 247
macro level   148, 154, 190
mail box   109
main memory   10, 13, 76
main store   13, 21, 302
mantissa dataflow   104
manufacturing verification test   317
mapping   171, 173

mask data generation 160
master image 204
MBA chip 108
memory bus 9, 16
memory control 77
memory identification 97
memory management unit 49
Memory Mapped I/O 304
microinstruction mode 21
microinstruction operation
  registers 20
microinstruction paging 11
microinstructions 11, 31
micromode 21, 25
MMU chip 12
mode 21
modulo 3 checking 106
Motorola 51, 112, 301
multiple clocks 181
multiple cycle paths 182
multiplier 105

N

NAND level 161
netlength 179
nets-of-agencies 190
noise 211, 220, 300
non-S/370 memory 11, 13, 21, 24, 308

O

off-chip nets 178
oscillator 18, 112, 119

P

packaging 227
page look ahead 27
page mode 83
page table 50
partition 158, 167, 179, 187, 285, 293
partitioned synthesis 167
Personalization 97
physical design 145, 187, 299
pico probe 261, 299
pin 165, 178

pipeline 24, 27
PLA 149, 155
placement 177, 222, 293
PL.8 311
poly-silicon 203
polyimide layer 230
power bus 207
power distribution 205, 249, 292
power on reset 135
power ring cell 292
power-on 316
prefetch buffer 10, 24, 41
primary inputs 247
primary outputs 247
probe cells 299
process parameter test 257
process tolerances 212
processor bus 9, 16, 56, 100, 108, 128, 140
processor bus commands 35
Processor card 142, 227
product assurance 316
program status word 55, 62, 322
programmable control space 307
punch through 268

R

random pattern testing 252
RAS 17, 132, 300, 304
real address 50
realisation 7, 318
redundancy steering 221
redundant bit 80, 93
redundant bit directory 80
reference bit 64
refresh 79, 86
regions 283
register-transfer-level 148, 190
regression test 323
reset 112, 115, 135
retry 92
REXX 155, 199
rounding bit 104
row access strobe 82

S

scenario file 160
scrub 79, 94
second metal test 259
segment table 50
sense amplifier 220
sense microinstruction 35
serial reset 138, 140
Service Processor 16, 134, 141, 301, 308
shadow approach 15, 99, 100, 106, 140
shapes 292, 293
shift register latch 116, 135, 142, 204, 246
shift unit 38, 104
signature analysis 252
simulation 313
simulation control 189, 198
simulation model 187, 193
simulator 190
sink partitions 168
soft errors 85
source partition 168
source stage 29
SQRT 15
SRL 116
STC bus 9, 16, 57, 75, 78, 140
STC chip 13, 77
sticky bit 104
storage key 64
store aligner 72
structural information 186
structural test patterns 245
stuck-at 85, 107, 244, 248, 259
subcircuit elements 208, 209, 213
support bus 9, 16, 18, 114, 141, 301
support bus adapter 141
synthesis experience 169
system bring-up 322, 325
system level 148, 155, 190
system level compiler 155
system measurement interface 18
system of models 186
system simulation 321
S/370 channel 108, 301
S/370 control space 306
S/370 mode 24
S/370 operation registers 20

T

target stage 30
technology data files 160
technology level 162
test cases 185, 188, 245, 320
test data generation 160, 245, 283
testing 222
thermal expansion 229
threshold voltage 269
time-of-day clock 44
timer 20, 44
timer control space 306
timing analysis 159, 177, 294
timing analyzer 175, 177, 181
timing correction 163, 168, 184
TLB 12, 40, 51
transconductance 269
transfer function 190
transforms 158, 160, 164, 171, 174
translation-lookaside buffer 12, 51, 55, 60, 64, 140, 215, 292
trap level 26, 33
traps 26
tree decoder 220
tristate driver 212, 248
true-complement switch 221

U

unit support interface 103, 114, 141, 142, 189

V

validity bit 64, 67
VAX 51, 99
VHDL 156
via holes 292
virtual address 50, 64
VLSI tester 243, 259

W

wireability 208
wiring 160, 177, 222, 293, 300
wiring capacitance 178
wiring channels 207, 208
wiring program 291
word decoders 220
write stage 30

Z

zero yield 249
zero-insertion-force 234